SAC Library

This book is due for return on or before the last date shown below.

27 MAY 1999
24 MAR 2000
26 JUN 2000
25 MAR 2002
15 APR 2002

29 JUL 2002
22 APR 2003
- 1 DEC 2003
12 FEB 2004

12 FEB 2009

07 NOV 2014

y

d

ts

Don Gresswell Ltd., London, N.21  Cat. No. 1208     DG 02242/71

AC 0000272 0

630 Bur

# Agricultural Sustainability
## Economic, Environmental and Statistical Considerations

Edited by

**Vic Barnett**
Rothamsted Experimental Station, UK

**Roger Payne**
Rothamsted Experimental Station, UK

and

**Roy Steiner**
Rockefeller Foundation, New York, USA

JOHN WILEY & SONS
Chichester · New York · Brisbane · Toronto · Singapore

Copyright © 1995 by John Wiley & Sons Ltd,
                   Baffins Lane, Chichester,
                   West Sussex PO19 1UD, England
                   National   Chichester (01243) 779777
                   International  (+44) 1243 779777

All rights reserved.

No part of this book may be reproduced by any means,
or transmitted, or translated into a machine language
without the written permission of the publisher.

*Cover illustration*   Photograph © Sean Ryan, Studio
108, Tel: (01249) 712399

*Other Wiley Editorial Offices*

John Wiley & Sons, Inc., 605 Third Avenue,
New York, NY 10158-0012, USA

Jacaranda Wiley Ltd, 33 Park Road, Milton,
Queensland 4064, Australia

John Wiley & Sons (Canada) Ltd, 22 Worcester Road,
Rexdale, Ontario M9W 1L1, Canada

John Wiley & Sons (SEA) Pte Ltd, 37 Jalan Pemimpin #05-04,
Block B, Union Industrial Building, Singapore 2057

**British Library Cataloguing in Publication Data**

A catalogue record for this book is available from the British Library

ISBN  0 471 95009 2

Typeset in 10/12pt Photina by Thomson Press (India) Ltd, New Delhi
Printed and bound in Great Britain by Biddles Ltd, Guildford and King's Lynn

# Contents

**Preface**     xi

**PART I   BASIC ISSUES**     1

1   **Agricultural Sustainability: Concepts and Conundrums**     3
    *R. W. Herdt and R. A. Steiner*

    1.1   THE STARTING POINT     3
    1.2   SUSTAINABILITY CONCEPTS     5
       1.2.1   Spatial levels     5
       1.2.2   Time     6
       1.2.3   Dimensions/forces     7
    1.3   SUSTAINABILITY MEASUREMENT     8
       1.3.1   Total social factor productivity     9
       1.3.2   Ecosystem health     10
    1.4   OVERVIEW     12

2   **Long-term Experiments and Their Choice for the Research Study**     15
    *R. A. Steiner*

    2.1   LONG-TERM EXPERIMENT SITES     15
    2.2   STATISTICAL DESIGN LIMITATIONS     17
    2.3   THE REPRESENTATIVENESS     17
    2.4   SOIL MOVEMENT     18
    2.5   RELEVANT TREATMENTS     19
    2.6   CHANGES IN NON-TREATMENT FACTORS     20
    2.7   DATE RELIABILITY     20
    2.8   ADEQUACY OF DATA COLLECTION     21
    2.9   DATA MANAGEMENT     21

3   **Economic and Statistical Considerations in the Measurement of Total Factor Productivity (TFP)**     23
    *A. I. Rayner and S. J. Welham*

    3.1   INTRODUCTION     23
    3.2   MEASURING THE PRODUCTIVE PERFORMANCE OF A SYSTEM OVER TIME     25

| | | |
|---|---|---|
| 3.3 | INDEX NUMBERS | 30 |
| 3.4 | MEASURING PROFITABILITY | 32 |
| 3.5 | TREND AND VARIABILITY IN TFP AND SUSTAINABILITY IN AN AGRONOMIC TRIAL | 33 |
| 3.6 | STATISTICAL ASSESSMENT OF TFP INDEXES | 35 |
| | 3.6.1 Estimation of index trend | 36 |
| | 3.6.2 Detecting a change in trend | 36 |
| 3.7 | CONCLUSIONS | 37 |

## PART II   DETAILED STUDIES                                                39

### 4   Long-term Cotton Productivity Under Organic, Chemical, and No Nitrogen Fertilizer Treatments, 1896 to 1992       41
*G. Traxler, J. Novak, C. C. Mitchell, Jr. and M. Runge*

| | | |
|---|---|---|
| 4.1 | HISTORIC PERSPECTIVE ON 'THE OLD ROTATION' | 42 |
| 4.2 | OLD ROTATION TREATMENTS | 42 |
| 4.3 | SITE CHARACTERISTICS AND SOIL MEASUREMENTS | 42 |
| 4.4 | PRODUCTIVITY ANALYSIS | 47 |
| 4.5 | OLD ROTATION DATA | 49 |
| 4.6 | RESULTS | 51 |
| | 4.6.1 Indexes without considering external costs | 42 |
| | 4.6.2 Total social factor productivity (TSFP) | 54 |
| | 4.6.3 Productivity effects of winter legume and chemical nitrogen | 57 |
| 4.7 | CONCLUSIONS | 60 |

### 5   Extrapolating Trends from Long-term Experiments to Farmers' Fields: The Case of Irrigated Rice Systems in Asia   63
*K. G. Cassman and P. L. Pingali*

| | | |
|---|---|---|
| 5.1 | INTRODUCTION | 63 |
| 5.2 | LINKING STRUCTURE TO FUNCTION OF CROPPING SYSTEMS | 64 |
| 5.3 | REGIONAL AND NATIONAL TRENDS IN PRODUCTIVITY AND YIELD | 65 |
| 5.4 | FARM-LEVEL TRENDS IN PRODUCTIVITY AND YIELD | 66 |
| 5.5 | YIELD TRENDS IN LONG-TERM EXPERIMENTS ON INTENSIVE RICE SYSTEMS | 71 |
| 5.6 | CAUSES OF THE YIELD DECLINE AND N FACTOR PRODUCTIVITY | 75 |
| 5.7 | LINKING TFP TRENDS AT THE EXPERIMENTS, FARM, AND REGIONAL LEVELS | 83 |
| | ACKNOWLEDGEMENTS | 84 |

### 6   Wheat/Fallow Systems in Semi-arid Regions of the Pacific NW America                                             85
*B. Duff, P. F. Rasmussen and R. W. Smiley*

| | | |
|---|---|---|
| 6.1 | INTRODUCTION | 85 |
| | 6.1.1 Location of the research center | 86 |

|   |   |   |   |
|---|---|---|---|
| | 6.2 | HISTORY OF THE RESIDUE MANAGEMENT EXPERIMENT | 86 |
| | | 6.2.1 Experimental design | 87 |
| | | 6.2.2 Fertilizer amendments | 87 |
| | | 6.2.3 Residue burning | 89 |
| | | 6.2.4 Adjustments and modifications | 89 |
| | | 6.2.5 Yield | 89 |
| | | 6.2.6 Soil analysis | 89 |
| | 6.3 | ANALYSIS AND INTERPRETATION OF RESULTS | 90 |
| | | 6.3.1 Yield history | 90 |
| | | 6.3.2 Soil carbon and nitrogen | 92 |
| | 6.4 | THE ECONOMIC ENVIRONMENT | 92 |
| | 6.5 | ESTIMATING TOTAL SOCIAL FACTOR PRODUCTIVITY | 99 |
| | | 6.5.1 Definitions | 99 |
| | | 6.5.2 Trends | 99 |
| | | 6.5.3 Influence of soil erosion | 104 |
| | | 6.5.4 Prices, profits and government programs | 105 |
| | 6.6 | IMPLICATIONS FOR BIOLOGICAL AND ECONOMIC SUSTAINABILITY OF WHEAT/FALLOW SYSTEMS | 107 |
| | 6.7 | SUMMARY | 107 |

## 7 Multi-crop Comparisons on Sanborn Fields Missouri, USA — 111
*J. R. Brown, D. D. Osburn, D. Redhage and C. J. Gantzer*

|   |   |   |
|---|---|---|
| 7.1 | INTRODUCTION | 111 |
| 7.2 | SELECTION OF STUDY MATERIALS | 112 |
| 7.3 | DESCRIPTION OF MANAGEMENT | 112 |
| 7.4 | PROCEDURES | 114 |
| | 7.4.1 Fertilisers | 114 |
| | 7.4.2 Soil characteristics | 114 |
| | 7.4.3 Soil erosion | 116 |
| | 7.4.4 Crop yields | 119 |
| | 7.4.5 Weather | 120 |
| 7.5 | PERFORMANCE OF EXPERIMENTAL PLOTS OVER TIME | 120 |
| 7.6 | AN INDEX APPROACH | 130 |
| 7.7 | CONCLUSIONS | 131 |

## 8 Major Cropping Systems in India — 133
*K. K. M. Nambiar*

|   |   |   |
|---|---|---|
| 8.1 | INTRODUCTION | 133 |
| 8.2 | CLIMATE | 134 |
| 8.3 | SOILS | 134 |
| 8.4 | EXPERIMENTAL DESIGN | 136 |
| 8.5 | SUSTAINABILITY ON ALLUVIAL SOIL AT LUDHIANA, PUNJAB | 137 |
| | 8.5.1 Yield trend | 138 |
| | 8.5.2 Soil reaction | 139 |
| | 8.5.3 Soil organic carbon | 139 |
| | 8.5.4 Available soil nitrogen | 140 |
| | 8.5.5 Available soil phosphorus | 141 |
| | 8.5.6 Available soil potassium | 141 |
| | 8.5.7 Available soil zinc | 142 |
| | 8.5.8 Discussion of the results at Ludhiana | 143 |

|     | 8.6   | SUSTAINABILITY ON MEDIUM BLACK SOIL AT JABALPUR, MADHYA PRADESH | 144 |
| --- | --- | --- | --- |
|     |     | 8.6.1 Yield trend | 144 |
|     |     | 8.6.2 Soil reaction | 146 |
|     |     | 8.6.3 Soil organic carbon | 146 |
|     |     | 8.6.4 Available soil nitrogen | 147 |
|     |     | 8.6.5 Available soil phosphorus | 147 |
|     |     | 8.6.6 Available soil potassium | 148 |
|     |     | 8.6.7 Available soil sulphur | 149 |
|     |     | 8.6.8 Discussion of the results at Jabalpur | 149 |
|     | 8.7 | SUSTAINABILITY ON LATERITE SOIL AT BHUBANESWAR, ORISSA | 150 |
|     |     | 8.7.1 Yield trend | 150 |
|     |     | 8.7.2 Soil reaction | 152 |
|     |     | 8.7.3 Soil organic carbon | 152 |
|     |     | 8.7.4 Available soil nitrogen | 153 |
|     |     | 8.7.5 Available soil phosphorus | 153 |
|     |     | 8.7.6 Available soil potassium | 154 |
|     |     | 8.7.7 Available soil sulphur | 155 |
|     |     | 8.7.8 Discussion of the results at Bhubaneswar | 155 |
|     | 8.8 | SUSTAINABILITY ON FOOTHILL SOIL AT PANTNAGAR, UTTAR PRADESH | 156 |
|     |     | 8.8.1 Yield trend | 156 |
|     |     | 8.8.2 Soil reaction | 158 |
|     |     | 8.8.3 Soil organic carbon | 159 |
|     |     | 8.8.4 Distribution of organic carbon in the soil profile | 160 |
|     |     | 8.8.5 Cation exchange capacity | 162 |
|     |     | 8.8.6 Available soil nitrogen | 162 |
|     |     | 8.8.7 Available soil phosphorus | 163 |
|     |     | 8.8.8 Available soil potassium | 164 |
|     |     | 8.8.9 Available soil sulphur | 165 |
|     |     | 8.8.10 Available soil zinc | 165 |
|     |     | 8.8.11 Discussion of the results at Pantnagar | 166 |
|     | 8.9 | CONCLUSION | 167 |
|     |     | 8.9.1 Gaps and future research needs | 168 |
| 9   | **Sustainability—the Rothamsted Experience** | | **171** |

V. Barnett, A. E. Johnston, S. Landau, R. W. Payne,
S. J. Welham and A. I. Rayner

|     | 9.1 | INTRODUCTION | 171 |
| --- | --- | --- | --- |
|     | 9.2 | OBJECTIVES OF THE ROTHAMSTED STUDY | 174 |
|     |     | 9.2.1 Park Grass | 175 |
|     |     | 9.2.2 Woburn continuous wheat | 176 |
|     |     | 9.2.3 Broadbalk | 177 |
|     | 9.3 | DATA COLLECTION: SOURCES, PROBLEMS AND METHODS | 179 |
|     |     | 9.3.1 Land | 180 |
|     |     | 9.3.2 Labour | 180 |
|     |     | 9.3.3 Horse and machinery | 182 |
|     |     | 9.3.4 Fertilizer, FYM and lime | 183 |
|     |     | 9.3.5 Seed | 184 |
|     |     | 9.3.6 Herbicide, fungicide and pesticide | 185 |
|     |     | 9.3.7 Yield | 185 |

|        |                                                          |     |
|--------|----------------------------------------------------------|-----|
| 9.4    | TFP INDEXES EXCLUDING EXTERNALITIES                      | 186 |
| 9.4.1  | The choice of a TFP index                                | 187 |
| 9.4.2  | Performance of experimental plots over time             | 191 |
| 9.4.3  | Comparison between plots                                 | 197 |
| 9.5    | TFP INDEXES INCLUDING EXTERNALITIES                      | 198 |
| 9.6    | DISCUSSION                                               | 206 |
|        | ACKNOWLEDGEMENTS                                         | 206 |

## PART III   REVIEW OF FINDINGS                                       207

### 10 Incorporating Externality Costs into Productivity Measures: A Case Study Using US Agriculture                     209
*R. A. Steiner, L. McLaughlin, P. Faeth and R. R. Janke*

|         |                                                       |     |
|---------|-------------------------------------------------------|-----|
| 10.1    | THE CHALLENGE OF EXTERNALITIES                        | 209 |
| 10.2    | PESTICIDE EXTERNALITIES                               | 210 |
| 10.2.1  | Regulatory costs                                      | 211 |
| 10.2.2  | Pesticides and human health                           | 215 |
| 10.2.3  | Environmental externalities of pesticide use          | 217 |
| 10.2.4  | Incorporating pesticide costs into the productivity analysis | 222 |
| 10.3    | FERTILIZER EXTERNALITIES                              | 224 |
| 10.3.1  | Fertilizer regulatory costs                           | 224 |
| 10.3.2  | Fertilizer health costs to society                    | 224 |
| 10.3.3  | Environmental costs of fertilizers                    | 225 |
| 10.3.4  | Incorporating pesticide costs into the productivity analysis | 226 |
| 10.4    | EXTERNALITIES ASSOCIATED WITH SOIL EROSION            | 226 |
| 10.4.1  | Off-site costs                                        | 226 |
| 10.4.2  | Positive externalities associated with soil quality enhancement | 227 |
| 10.5    | CONCLUSIONS                                           | 229 |

### 11 Long-term Experiments and Productivity Indexes to Evaluate the Sustainability of Cropping Systems                 231
*K. G. Cassman, R. Steiner and A. E. Johnston*

|      |                                                          |     |
|------|----------------------------------------------------------|-----|
| 11.1 | MAINTAINING RELEVANCE OF LONG-TERM EXPERIMENTS           | 232 |
| 11.2 | QUANTITATIVE MEASURES OF SUSTAINABILITY IN LONG-TERM EXPERIMENTS | 236 |
| 11.3 | SYSTEM PERFORMANCE AND SOIL QUALITY                      | 239 |
| 11.4 | RECOMMENDATIONS FOR CONDUCTING LONG-TERM EXPERIMENTS AND ANALYSIS OF PRODICTIVITY TRENDS | 241 |
| 11.5 | CONCLUDING REMARKS                                       | 243 |

**References**                                                         245

**Index**                                                              257

# *Preface*

**Sustainability** has become an important concern for environmentalists, agricultural researchers, farmers, politicians and ordinary voters. All agree on the importance of ensuring that agricultural systems of production can, in some sense, be sustained, i.e. be operated in such a way that quality of output can be maintained year after year, without degradation of the environment. There is less agreement, however, on how this concept of *sustainability* should be interpreted, or precisely defined, let alone measured. One important aspect concerns the need to devise good practices of farming to ensure that fertility is maintained and that insect and fungal pathogens can be controlled. However, these cannot be assessed without considering the economic viability of the suggested systems within a farm. Nor should they be assessed without investigating the wider environmental effects, for example of pesticides on wildlife or fertilisers on water supplies, but these *externalities* can be very difficult to measure. The subject thus presents challenges for scientists from many disciplines including agriculture, biology, economics, nutrition and statistics.

By the very nature of the topic, new experimental research into sustainability cannot be expected to yield results in the short term. A rich vein of information, however, is available in the long-term experiments and investigations that have been conducted over many years—150 years in the case of Rothamsted. The experience of those involved demonstrates how it has been possible to maintain crop yields under some very diverse conditions. Others, however, have encountered difficulties and some experiments have had to be discontinued.

It was against the joint background of lack of precision in defining the sustainability concept and the wealth of potential information in the long-term experiments, that the Rockefeller Foundation commissioned an important international research study on sustainability. They invited six major centres of long-term experiments (in India, the Philippines, the UK and the US) to conduct detailed studies of their data and to address the issue of definition and, in particular, to make proposals on *how to measure sustainability in quantitative terms*. The chosen long-term experiments were:

- Classical Experiments, Rothamsted Experimental Station, UK (started 1843)
- Sanborn Fields, University of Missouri, USA (1888)

- Old Rotation Study, University of Auburn, Alabama, USA (1896)
- Residue Management Experiment, University of Oregon, USA (1931)
- Long-term Continuous Rice Cropping Experiment, Los Banos, Phillipines (1963)
- All-India Coordinated Long-Term Fertilizer Experiments, 11 sites in India (1972).

Each centre produced a detailed report, and an international workshop on the theme *Measuring Sustainability Using Long-Term Experiments* took place at Rothamsted Experimental Station from 28–30 March 1993, to discuss the results. This meeting provided the culmination of the research project sponsored by the Rockefeller Foundation, in recognition of the importance of this topic, not merely for current research but also for future generations.

This book draws together the conclusions of the six separate studies of the workshop. It is compiled and edited by three of the principal workers on the joint research project, with each chapter written by researchers who carried out the work. Part I (Chapters 1–3) considers basic issues of definition and the choice of appropriate economic and statistical methodology. The extensive Part II (Chapters 4–9) presents the results of the six individual research studies, whilst Part III (Chapters 10–11) draws together the conclusions, addresses the thorny issue of environmental externalities and makes proposals for future research and implementation.

The book has been structured as a comprehensive and unified contribution to the important issue of sustainability and we hope that it will provide a valuable resource for further study.

**V. Barnett**
**R. W. Payne**
**R. A. Steiner**
*April 1994*

# PART I

*Basic Issues*

# 1
# *Agricultural Sustainability: Concepts and Conundrums*

R. W. Herdt and R. A. Steiner

*Rockefeller Foundation, NY, USA*

Sustainable development is development that meets the needs of the present without compromising the ability of future generations to meet their own needs. (The World Commission on Environment and Development, 1987).

'Sustainability' is the buzz word of the decade, invoked by presidents, priests, and plebeians. Agriculturists have chanted the sustainability refrain with as much vigour as any, but few have ventured to define the concept clearly; even fewer to suggest how it may be measured (Conway and Barbier, 1990; Hart and Sands, 1991; Norguard, 1991; Ehui and Spencer, 1993). The work reported in this book was stimulated by a desire to move beyond the rhetoric that characterizes much of the sustainability debate and operationalize the concept.

## 1.1 THE STARTING POINT

All must be concerned with future food availability. Global population has more than doubled from around 2.5 billion in 1950 to over 5.3 billion in 1990; even with the most optimistic projections for adoption of family planning, it is expected at least to double again before stabilizing. Agriculture produces essentially all the food in today's world. Ocean fish contribute the only significant contribution to

---

*Agricultural Sustainability: Economic, Environmental and Statistical Considerations.*
Edited by V. Barnett, R. Payne and R. Steiner © 1995 John Wiley & Sons Ltd

global food from 'wild' sources. Even exotic items like the wild rice gathered by North American tribes, and fruit drinks made from Brazil's forest trees, have become domesticated and are now produced by deliberate cultivation (i.e. agriculture). 'High tech' food production using biotechnology in factory-like settings, if ever it becomes economically competitive, will require 'feed stock' produced on the land, and an energy source, for which the sun is likely to be the most practical. Thus, food production may look very different from today, but agriculture of some form will have to produce the food needed to feed a doubled population in the future.

Sustainable agriculture is concerned with the ability of agricultural systems to remain productive in the long run. The studies reported in this book take a quite straightforward approach to determining whether particular crop production systems are productive: they look at systems that have been in operation for a long period on long-term experimental research stations, determine what has happened to their productivity over the period, and ask what the record suggests about their sustainability into the future.

In the view of some, 'Compelling physical evidence from around the world suggests that current framing practices in many areas cannot be sustained much longer. Agricultural resources are threatened or declining' (Faeth, 1993). However, in the aggregate, per hectare agricultural production has been increasing over time and more than keeping pace with population growth. Hence, some ask, what evidence suggests that current farming systems cannot be sustained? The two views appear to be inconsistent, but because one refers to particular farm fields and the other to regional or national aggregations, they are not necessarily inconsistent. It is important to know whether the present aggregate situation will continue to dominate into the future, or whether we face a future of declining productivity. This question has stimulated the present effort to examine carefully whether long-term trends in productivity provide evidence for the decline some claim is widespread.

Any effort to answer the sustainability question through an appeal to farmers' collective experience is confounded by the continuous increase in inputs applied in most farming systems, which increase yield and may offset reductions in the underlying productive capacity. Because many inputs are produced off the farm and often require petroleum products, some people believe that in the long run it will be impossible to continue to rely on their use. Thus, the important question is not whether agricultural productivity is changing. It is whether agricultural productivity gains are occurring at the cost of degradation in the underlying resource base which will eventually result in falling productivity. Answers to that question could, conceptually, be explored by examining changes in productivity and the resource base in farming systems where inputs have been held constant over time. Unfortunately, there are virtually no data that permit such a joint examination on a national or other large-scale basis. On the scale of farm fields, or more precisely, experimental station fields, however, there are many crop production trials that have been conducted over periods of 25, 50, or even 100 years or more. These seem a natural place to look for answers to the sustainability question. Thus, in this book the authors report the results of their efforts to examine crop production systems under well defined circumstances where good records on the types and costs of inputs have been kept and where the inherent productive capacity of the resource base has been monitored over long periods of time.

Aside from the potential difficulties in finding data to determine whether production practices are sustainable, there is no agreed *definition* of what sustainability means. As everyone who has contemplated it knows, sustainability is a complex idea involving many different aspects of human activity and linked to many levels and dimensions of the global system. A brief discussion of some relevant concepts and of our definition of sustainable agricultural production is, therefore, warranted.

## 1.2 SUSTAINABILITY CONCEPTS

Agricultural sustainability is concerned with production of agriculture over time. Lynam and Herdt (1989) argue that sustainability must be defined with respect to systems, rather than inputs or crops, because crop varieties and inputs produce nothing in isolation. Only when combined as components of a system do they produce output; the question in whether or not the system of production is sustainable. Sustainability is the result of the relationship between technologies, inputs and management, used on a particular resource base within a given socio-economic context. Careful recognition of three aspects of systems—*space, time,* and what we call *dimension*—help make the discussion concrete.

### 1.2.1 Spatial levels

One may consider systems across an infinite range of space: global, regional, farm field, individual plants, and microscopic. Figure 1.1 illustrates how, as one proceeds from the global to the local level, the number of types of system increases rapidly. This leads to some of the difficulty inherent in the sustainability concept, and reinforces the need to define carefully the spatial dimension. The number of levels, and their interconnections, are part of the problem of determining when sustainability is an inherent property of a given system, and when sustainability is so dependent on external forces that it can be most usefully examined in a higher level system (Lynam and Herdt, 1989).

Simply defining the level of a system under consideration is only a beginning because at any level (except the universe) many possible types of system exist. Thus a clear definition of the particular system of interest is necessary in order to distinguish it from other types at the same level. In agriculture, one can more easily *look at* a cropping system of interest on a particular field than *define* it

## 6  Sustainability: concepts and conundrums

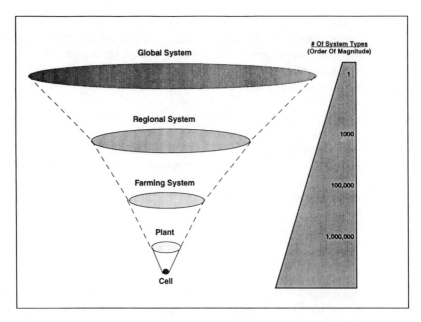

**Figure 1.1**  Spatial levels and associated number of systems.

unambiguously, because crop and animal production is affected not only by the weather, with its variation in rainfall, solar radiation and temperature, but also by the biotic environment in which insects, diseases, microbes and fungi affect the system, and by the soil with its physical, chemical, and biological characteristics. Care is necessary to define a system adequately, and the journalistic way in which systems are often described is entirely inadequate for analytical purposes. In some sense, each system is unique, or if not unique, only somewhat representative of a class of closely related other systems. This also makes it more appropriate to consider the sustainability of specific fields rather than poorly described systems.

### 1.2.2  Time

Sustainability can only be thought of in the context of a defined time period. Physics has shown that the sun is a dying star, hence all processes dependent on the sun are strictly not sustainable, but the time period over which the sun is dying is so long as to be meaningless for agriculture. At the other extreme, many seeds will germinate and grow briefly without being planted, but without, sunlight, water, and nutrients, growth will not continue for more than a few days: an isolated seed is not sustainable beyond a brief period.

Consideration of the time dimension is further complicated by the dynamic nature of reality. The empirical studies reported in this volume illustrate that in the real world agricultural production systems are constantly changing. Almost everywhere the systems in use by farmers today are different from those used 25 years ago. Natural variation and the influence of highly variable primary forces such as weather and technology render most short-term perspectives meaningless. Anticipating some of the empirical work examined in later chapters, experience shows that in most cases the important trends affecting the sustainability of a system usually become apparent in the first 20–40 years. In this work we are concerned with the sustainability of systems that have been in existence for at least 20 years.

### 1.2.3 Dimensions/Forces

Beyond the difficulty of defining system boundaries and specifying the time period, measuring sustainability is further complicated by the dimensions in which people think about the human condition. Plant growth is a biological process that results in physical changes, but agriculture is an economic activity serving a social purpose. The three dimensions, biological, economic, and social, may be analysed in very different, although interrelated ways.

**Biological/physical** The biological or physical dimension can be reflected in the quantity of output, which depends on the physical quantity of inputs and the biological growth processes. Degradation of the resource base can lead to falling quantity of output over time, caused by erosion, waterlogging, the destruction of soil structure, and similar physical or biological phenomena. On different levels, changing climate or the introduction of new plant diseases and pests can have a similar effect.

**Economic** The economic dimension can be reflected in the value of output, i.e. the quantity times a price that represent value. Even if the quantity of a system's output is constant over time, the economic environment can lead to the system's failure because of falling commodity prices, increased input costs or other economic changes.

**Social** The social dimension may be reflected in the capacity of systems to adequately support farming communities. Agricultural systems depend on human communities and institutions. When these communities deteriorate agricultural production may fall. Poor agricultural policy, insecure land tenure, war, social disruption, and changing labour conditions are only a few social factors that can lead to non-sustainable agricultural systems. The biological/physical sustainability of an agricultural system will affect is capacity to sustainably support a social system.

All three dimensions of sustainability are important, and one might think of them as being interrelated as in Figure 1.2. Some observers focus on the physical/biological aspect of sustainability, others on the social or economic dimension.

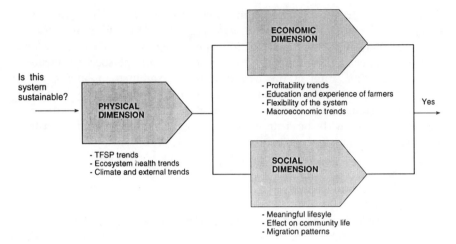

**Figure 1.2** Questions within the three dimensions of sustainable agriculture.

While economic performance has important effects on farmer's decisions, biological performance (the amount and kind of plant growth generated by a system) is the base which is modified by costs and prices to give economic performance. Social performance is dependent on both physical and economic performance.

Even confining one's view to the physical/biological dimension, some observers focus on performance of individual fields, others of farms or villages, others of regions, and still others of national agricultural systems. Regional and national performance is the result of many farmers' activities, so if one understands what is happening on farm fields, then we believe one has a better basis for understanding what is happening at the aggregate level.

Our interest here is in the sustainability of cropping systems on farm fields over periods from 20 years to over 100 years, and with agronomic or biological sustainability as well as economic sustainability. Agricultural researchers concerned with ensuring that technologies arising from crop research do not exacerbate the deterioration of systems must understand this larger view of sustainability before they focus on the practical steps needed to apply the concept to their work. But the practical steps require a concrete way of *measuring sustainability*, not just a set of concepts.

## 1.3 SUSTAINABILITY MEASUREMENT

Figure 1.3 depicts a crop system on a farm field. Inputs are applied to the land and outputs are produced. The most frequently employed measure of agricultural productivity is yield, generally expressed as grain yield per hectare per crop. In most cases this is an inadequate measure of sustainability because it ignores time,

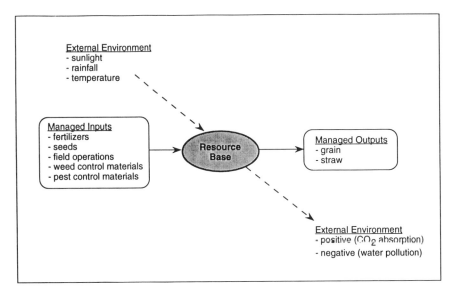

**Figure 1.3** Input–output model of a cropping system

ignores secondary products like straw, ignores input use other than land, and ignores externalities, all of which should be included in a sustainability measurement.

### 1.3.1 Total social factor productivity

Inputs are relevant to sustainability: it would be difficult to agree that a system which requires increasing inputs to maintain constant outputs over time is sustainable. *Total factor productivity*, which measures an index of total output relative to an index of total inputs is a better measure of productivity than yield because it recognizes the use of all inputs, including seeds, water, labour, traction, pesticides, fertilizers, microorganisms, and others. In terms of Figure 1.3, total factor productivity reflects the relationship between outputs and managed inputs.

However, total factor productivity as usually measured does not take account of inputs and outputs that are *external* to the production decision-making units. Some inputs are managed, that is they are provided by those decision-makers whose behaviour is intended to control the system of interest. Outputs are usually taken to be the desired yields, but production may result in some undesired outputs (ground-water pollution and insecticide drift, for example) or even unrecognized products that move off the land. Economists call these latter quantities *externalities* because they relate to economic units external to those who decided on the inputs and who receive the output from a production process.

External outputs of one economic unit may become external inputs to another, for example acid rain generated by industrial companies.

*Total social factor productivity* (TSFP) has been defined as being more inclusive than total factor productivity, accounting not only for the managed inputs but for the externalities as well (Herdt and Lynam, 1992). That is, outputs include those obtained on the site, such as grain and erosion of fields, and those that are external to the site, such as water pollution, change in pastoral scenery and pesticide residues.

To measure sustainability using this approach it is necessary to measure the flow of inputs and outputs across the boundaries of the system over time, taking care to include all the inputs and outputs, and to aggregate each set so as to produce a single measure of input and a single measure of output. The problem of combining many different kinds of input and output into a single measure, well known to economists as the index number problem, is discussed in Chapter 3. At this stage, it is sufficient to note that various inputs must be converted to some common unit of measurement. Among the measures that have been used are prices, energy equivalents, and carbon equivalents.

Because most of the long-term experiments examined in this project have well controlled if not constant inputs, yield is closely correlated with the TSFP index. However, on typical farms this would not necessarily be the case because inputs are constantly being adjusted.

### 1.3.2 Ecosystem health

Even given a good measure of all inputs and all outputs, it is also necessary to know what is happening to the underlying quality of the resource base since, by definition, the quality of the resource base is expected to affect future physical/biological production. Figure 1.4 illustrates resources base characteristics that contribute to expected future productivity and may change over time.

Consider soil organic matter, one aspect of soil fertility. If a field begins with a high level of soil organic matter and produces constant or increasing output using constant inputs over time, but at the end of the period has a severely depleted soil organic matter status, can the production system be considered sustainable? Likewise with soil phosphorus. Crop yield may be relatively insensitive to levels of phosphorus over a wide range of available phosphorus, so a system may provide non-declining total factor social productivity while at the same time depleting the phosphorus. When the level is depleted below a critical value, yields will fall, and it may take more than mere application of the amount 'normally required' by a crop to return yields to their former level. Taking account of the 'consumption' of soil phosphorus over time will give a better indication of the sustainability of a system than ignoring it.

'Ecosystem health' can be used as a measure of sustainability only if a good criterion for resource base quality can be identified and measured. Good indicators

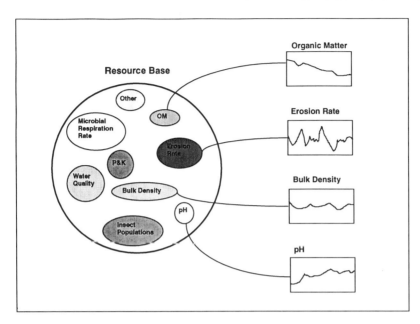

**Figure 1.4** Conceptual framework of ecosystem health indicators

are those which reflect whether a system will be productive over a long period. In other words, good indicators of ecosystem health are good indicators of sustainability! This circularity can be broken only by determining from experience the conditions that are associated with systems that maintain their productivity and the conditions associated with unproductive systems.

Although there is no universally accepted set of ecosystem health measures, some, mostly reflecting various soil propeties, are broadly accepted. Soil organic matter, soil water quality, soil structure, soil micro-organisms, pH, and soil temperature may all change as a consequence of the inputs applied and the outputs removed. Erosion is also important, but it can be reflected as an 'output' from the system, albeit a negatively valued output. The prevalence of pest and beneficial organisms in the environment may be another indicator of ecosystem health.

One difficulty in applying such criteria is that what is and is not healthy depends on the ecosystem and the agricultural system. There are a number of criteria, such as excessive erosion, which are always bad for a system. However, many criteria may be good for one cropping system and not for another, so the health of the resource base must be measured for a particular cropping system on a particular field. Then, even though one can devise productive systems for a wide range of pH values, the plants in the cropping gsystem of interest will prefer some range of pH. Movement towards either a more acid or more alkaline condition would indicate a deterioration in the resource base for the system.

## 12  Sustainability: concepts and conundrums

Another problem that arises in empirical applications is that some indicators of system health may be positive while others are negative, making a clear conclusion (or diagnosis) impossible (Altieri, 1986). It may be possible to develop a soil health index, but at this time there is not sufficient knowledge to be sure of its validity. Because of these difficulties, the case studies that follow do not focus on the system health approach to measuring sustainability. However, this aspect of sustainability clearly requires further attention if valid conclusions are to be made.

## 1.4  OVERVIEW

Figure 1.5 summarizes the set of questions that may be asked to determine whether a particular system is physically/biologically sustainable, using both the total social factor productivity and the ecosystem health criteria. A *non-decreasing trend* in yield and in total social factor productivity is necessary to call a system sustainable. In addition, the ecosystem health indicators should remain at acceptable values. If the measurement problems can be solved, both the ecosystem health and the output/input aspects can be evaluated for any system. If

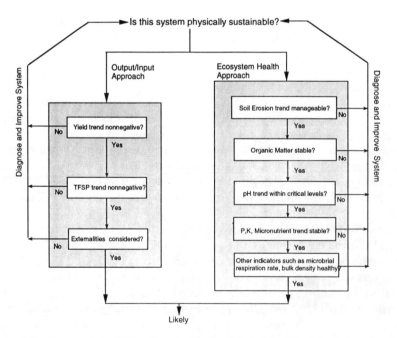

**Figure 1.5**  The sustainability flowchart: is this agricultural system physically sustainable?

the two approaches are equally valid and consistent, any ecosystem that is 'healthy' should have a non-decreasing trend in the TSFP, and vice versa. In reality, many variables in both approaches are difficult to identify and measure. We also suspect that, since the linkage between the conditions that reflect ecosystem health and the output/input performance is poorly understood, empirical application may result in one approach indicating sustainability whilst another does not.

Because sustainability is about the future, both measurement approaches require prognostication of what the observations mean for future productivity. The TSFP ratio implicitly assumes that past trends are a good indicator of future trends, the other that ecosystem health trends are a good indication of future performance.

The definitions provided above help to clarify some basic considerations in the empirical measurement of sustainability. In later chapters each of the authors explores various approaches to measuring the sustainability of their respective cropping systems using these considerations. As expected there is a rich variety of experience that significantly advances our understanding of how to go about measuring sustainability Among the many questions that came up in the course of analysing the data were:

(a) How long a time trend is needed to conclude something about sustainability?
(b) How can changing technology be reflected in a consistent manner?
(c) How should natural variation be treated?
(d) What are the relevant externalities and how can they be measured?
(e) Which measure or combination of measures will give the clearest picture of the sustainability of farming system?

What quickly becomes clear is that even when restricting the examination to the easiest dimension of sustainability—the physical/biological—the problem is fraught with complications. Resolving these complications, and incorporating the social and economic dimensions of sustainability will, no doubt, provide numerous challenges for years to come.

# 2
# Long-term Experiments and Their Choice for the Research Study

R. A. Steiner

*Rockefeller Foundation, NY, USA*

The most convincing proof of the sustainability of any agricultural system is a long-term experiment with positive results. Even with their limitations, long-term experiments provide the only reasonable empirical base on which we can evaluate the concept of sustainability. Because agriculture is location-specific, no group of experiments can adequately represent the agricultural systems that feed the world, and the experiments chosen here in no way claim to do so. However, the experiments described in this study are illustrative of the diversity of experiences that are typical of any group of agricultural systems. Through these experiments, one can observe the tremendous variability within the systems themselves as well as productivity trends that range from clearly positive to distinctly negative.

## 2.1 LONG-TERM EXPERIMENT SITES

At the outset of this research, limited information was available about the existing number of long-term experiments. Agricultural experts from the World Bank and FAO estimated that the total number of such experiments in the world was between 20 and 30. The scientists who collaborated with this project were from this original list of 30 known experiments. However, one of the most valuable and

---

*Agricultural Sustainability: Economic, Environmental and Statistical Considerations.*
Edited by V. Barnett, R. Payne and R. Steiner © 1995 John Wiley & Sons Ltd

**Table 2.1** Geographical breakdown of long-term agricultural experiments

| Region | Number of long-term experiments over 10 years old |
|---|---|
| Africa | 15 |
| Australia | 12 |
| North America | 56 |
| Former Soviet Union | >100 |
| Asia | 48 |
| South America | 17 |
| Europe | >100 |

unanticipated results of this study was the discovery of literally hundreds of often unknown and poorly documented long-term experiments around the globe. These have recently been compiled into a global directory which will be available shortly from the FAO (Steiner and Herdt, 1995). Their geographical breakdown is illustrated in Table 2.1.

The criteria for including experimental sites in this collaborative study were:

- The length of the experiment (as long as possible).
- The reliability and organization of the data.
- The extent to which the experiments reflected a diversity of crops and agroecological environments.
- Researchers' willingness to participate.

After sending our surveys to all known experiments, researchers responsible for the following experiments agreed to collaborate:

- The Rothamsted Classical Experiments, United Kingdom: started 1843.
- The Sanborn Fields, Missouri, U.S.A: started 1888.
- The Old Rotation Study, Auburn, Alabama, USA: started 1896.
- The Residue Management Experiment, Oregon, USA: started 1931.
- Long-Term Continuous Rice Cropping Experiment, Los Banos, Phillipines: started 1963.
- All-India Co-ordinated Long-Term Fertilizer Experiments, 11 sites in India: started 1972.

All these sites house a number of long-term experiments, each of which has many treatments. Because the primary purpose of this study was to evaluate various measures of physical sustainability, the analysis covers only a small number of these treatments. In most cases, the control treatment, the full NPK treatment and some alternative organic amendment treatments were analysed, but there are variations to this regime.

Because of the exploratory nature of this project some selection criteria were not given enough emphasis. For example no long-term experiment is static; they all undergo major changes at various points in their life as agriculture evolves. Thus for the purposes of evaluating the sustainability of particular systems, 30 and 40 year old experiments can be as valuable as older ones, and do not suffer from many of the limitations older experiments face such as soil movement and poor statistical design. However, if these experiments are to be used effectively, their limitations must be understood and addressed. Let us consider some of the more important limitations.

## 2.2 STATISTICAL DESIGN LIMITATIONS

Many of the classical long-term experiments were initiated before statistical theory and methods were developed, and thus many experiments which began before 1930 are not randomized and properly replicated. In fact, the Rothamsted classical experiments were highly influential in the development of modern statistical experimental design theory (Barnett, 1994). At a minimum one needs at least two replications to make statistically valid observations about treatments. Without these replications one cannot be sure if the differences observed are the result of other variables such as soil and micro-climate, or the treatments applied. However, even though data from such studies may not be analysable using conventional statistical methods, the results can still be used to make statements about the long-term, obvious effects of treatment practices. Furthermore, repetition over many years lends a certain level of confidence to conclusions about treatment differences (Frye and Thomas, 1991). In a sense, years substitute for replications. Where there are no replications, the level of confidence in any conclusion will depend on the particular circumstances of that experiment.

Another major flaw in the design of many long-term experiments is that the original plot size was too small. As a result, when the plot was split later in its life, the net plot, which should be about half the width of the full plot to eliminate border effects, almost disappeared. For example, the Old Permanent Manurial experiments in India began with plots that were originally $20\,m^2$ but now have been split and reduced to about $8\,m^2$ which leaves a net plot of $2\,m^2$. In this situation one can have little confidence in the results because of border effects and soil movement. In contrast the plots on the Rothamsted 'classicals' have remained of good size over the 150 years of their use.

## 2.3 THE REPRESENTATIVENESS

In most cases, experimental stations are located in the most productive agricultural land with limited slope and fertile soil. As such they are not representative of average cropping conditions, and conclusions cannot be extrapolated to more marginal areas. Perhaps one of the reasons there are so few long-term experiments in marginal areas is that they were simply unsustainable.

Another factor that limits site representativeness is that even though long-term experiments were initially located in rural areas, they are now surrounded by the cities and towns that were once far away. Such experiments as the Sanborn fields, the Aomori Experiment Station in Japan and the Morrow plots in the United States are essentially located in urban settings complete with traffic, smog and street lights. Besides the obvious disruption of such an environment, experiments located in such settings are much more prone to vandalism and the hazards of such things as drunken drivers and student pranks. Unless they have been designated as historical sites, the pressure to use the land for more 'valuable' purposes can be great and many experiments have been terminated to make way for a building site or parking lot. One experiment in the US, the McGruder plots, was physically lifted up and transported a few kilometres away because the university wanted to expand some buildings. In another case (the Sanborn Fields) the university eliminated several stories of a proposed building to avoid shading the long-term experiments. Needless to say, these transformations of the surrounding environment make the experiments quite different from those on farms in completely rural areas.

Just as the soil resources and surrounding environment of the experimental site can be significantly different from the average farm, the management practices also tend to differ dramatically. A different level of care and attention is given to experimental plots, and intensive management practices are the norm rather than the exception. Some of the management differences can be traced to the fact that reseachers work five days of the week whereas farmers work all seven. As a result, the researchers choose management practices that fit their schedule rather than fitting their schedule to the crop management needs.

## 2.4 SOIL MOVEMENT

Plot-to-plot soil movement caused by tillage, soil fauna, wind and water is a serious problem which gets worse with time. Sibbesen (1986), using soil movement models fitted to soil-phosphorus recordings from two 90-year old field experiments, estimated that in 21 of the world's field experiments which are more than 50 years old, only 28% of the original plough layer soil remains in their net plots today. This raises serious questions about the accuracy of any conclusions. The problem might have been avoided if the plot sizes had been large enough or grass borders had separated the plots. The size needed to ensure that there is not too much mixing of the soil between net plots without borders is given in Table 2.2. As one can see, very few long-term experiments have plots large enough to prevent soil movement affecting their results.

**Table 2.2** The smallest plot width that can be accepted in long-term field experiments when 95 and 99% of the total net addition of substance would remain in the net plot when the experiments terminate. (Assumptions are: constant annual net addition of substance; constant annual tillage; same $D$-value (transport coefficient of the soil) for both dimensions; square plots; width of the net plot is half the width of the treated plot.). Adapted from Sibbesen (1986)

| $D$ (Transport Coefficient) ($m^2$ year$^{-1}$) | Duration of Experiment (years) | Percentage of total net-addition of substance remaining in the net plot | |
|---|---|---|---|
| | | 95% | 99% |
| | | Minimum plot width (m) | |
| 0.1 | 10 | 7 | 10 |
| | 25 | 10 | 14 |
| | 50 | 14 | 20 |
| | 100 | 19 | 28 |
| 0.2 | 10 | 9 | 13 |
| | 25 | 14 | 20 |
| | 50 | 19 | 28 |
| | 100 | 27 | 39 |
| 0.3 | 10 | 11 | 16 |
| | 25 | 17 | 24 |
| | 50 | 24 | 34 |
| | 100 | 27 | 39 |

## 2.5 RELEVANT TREATMENTS

One of the greatest challenges facing all long-term experiments is maintaining their relevance for current agricultural practices. Almost all long-term experiments have been modified to some extent to reflect changes in prevailing farming practices with regard to plant populations, varieties, fertilizers and field operations. However, a number of experiments still maintain treatments that have ceased to be viable as cropping systems in their area. For example, one US experiments trucks in their manure from over a hundred miles away since animal operations have disappeared from the area, and another experiment uses pea vines as an organic amendment even though this treatment ceased to be a common practice over thirty years ago. Even if outside practices have changed, one can still make a case for maintaining a treatment (e.g. a manure treatment) if it contributes to an understanding of the system that could not be obtained otherwise. When changes do occur one has to question to what extent long-term continuity, the main asset of such trials, has been lost.

Experiments that began with large plots have numerous advantages because the plots can be split the accommodate new technologies. Where this option is not available, changes have to be well thought out, or the experiments will risk losing consistency.

## 2.6 CHANGES IN NON-TREATMENT FACTORS

Although treatments may remain the same from year to year, non-treatment factors such as weed and pest control practices, cultivation practices, and crop varieties vary dramatically. Agricultural systems are very complex because of the high degree of interrelatedness among their components. Experiments should (but usually do not) document changes in non-treatment factors, and thus fundamental forces affecting the system are ignored. For example the pesticide treatments in the long-term cotton experiments of Alabama have varied tremendously over its life, varying from no pest control, to DDT, to the latest integrated pest management. These pest control strategies have had a much more profound effect on yields than the soil amendment treatments being studied, but are almost ignored. Another fundamental change that all experiments over thirty years old have experienced is in the varieties grown following the green revolution. One can argue that these new varieties are fundamentally different systems because of the way they partition biomass and that any analysis should split the time data into pre-green revolution and post-green revolution systems.

## 2.7 DATA RELIABILITY

Given the nature of long-term experiments which often span the lifetime of many researchers, there is bound to be some variation in data collection methods and the types of data collected. The consistency of the data should be studied carefully. Many long-term experiments have experienced some type of accident or mistake, whether it is the application of amendments to the wrong plot, or the loss of data books. Given the length of the experiments this is unavoidable and usually not fatal, but it does leave gaps. The effect of such mistakes can also persist for a surprisingly long time. One experimenter mistakenly put manure on the control plot and, even though the manure was hand-picked off the field before it was ploughed, that plot showed increased yields for the next five years.

The methods used to measure various soil characteristics and yield should also be consistent from year to year. Where they are not, one must make compensating adjustments to allow for differences due to method. One of the problems in recording yield data is in the definition of net-plot changes over time. In the long-term experiments in Coimbatore, yield measurements are made using the entire plot, half the plot and everything in between, depending on the year and the researcher. Where there is no consistency in data collection techniques one can have little confidence in the results. This is why soil archives are such valuable assets. Most long-term experiments unfortunately do not have such archives, and this greatly diminishes their value. Many experiments had a soil archive initially but over time they were thrown out because of poor record keeping or the lack of foresight of space-conscious administrators.

## 2.8 ADEQUACY OF DATA COLLECTION

A major limitation of most long-term experiments is that many important types of data have not been collected limiting their use for such objectives as validating soil and crop models over time. As a minimum one should collect the following soil characteristics.

**Minimum data set**
Yield measurements using a consistent net plot
All inputs used
Weather data
Baseline description of soil, preferably archived

*Chemical/Mineral Measurements*
 – total carbon, OM
 – pH
 – P and K levels
 – relevant micro nutrients

*Physical Measurements*
 – soil erosion
 – water infiltration rate
 – field bulk density (under standard conditions)

*Biological Measurements*
 – microbial respiration rate
 – soil enzyme activity.

## 2.9 DATA MANAGEMENT

Most long-term experiments manage data poorly. Only recently have they begun to consider computerizing their data storage systems; this has low priority on many experiment stations since data organization can always be done next year! As a result, these data sets are extremely difficult to analyse. Some data sets are also poorly documented. Over the years different notations and weights are used and, unless the original researcher is there to explain what happened, data interpretation becomes a matter of guesswork.

Despite these limitations, the long-term experiments described in this study remain an invaluable resource with which to evaluate the sustainability of agriculture. Each experiment provides a picture of how agricultural systems in that part of the world have affected their physical environment and how the ability to provide food has varied over time.

# 3
# Economic and Statistical Considerations in the Measurement of Total Factor Productivity (TFP)

**A. I. Rayner[†] and S. J. Welham[‡]**

[†] *University of Nottingham, UK*
[‡] *Rothamsted Experimental Station, UK*

## 3.1 INTRODUCTION

In the past decade or so, concerns over the sustainability of modern agricultural systems have become a major item on the agenda of the agricultural research community. At issue is the capacity to sustain growth in agricultural production in the face of resource and environmental constraints. Several explanations have been put forward for the current emphasis on sustainability, see Ruttan (1991) and Graham-Tomasi (1991):

(i) The growth of population and income are placing further demands on agricultural systems which have limited opportunities for bringing new land into production. Most of the incremental production to meet additional demand must come from higher yields. Investment can also make some currently marginal land usable but the fragility of and environmental values associated with such land might impose constraints on production potential.
(ii) In the past and especially since World War II, the application of science to agriculture has resulted in productivity gains which have more than

---

*Agricultural Sustainability: Economic, Environmental and Statistical Considerations.*
Edited by V. Barnett, R. Payne and R. Steiner © 1995 John Wiley & Sons Ltd

compensated for the rapid growth in demand. However, further incremental production responses to inputs may be more difficult to obtain in the future. Physiological limits to improvements in crop yields and livestock efficiency could retard productivity gains stemming from advances in conventional technology. Developments in molecular biology and genetic engineering could release these constraints but general adoption of agricultural practices incorporating the new technology is some way into the future.

(iii) Science based agriculture relies heavily on external inputs—fuels, chemicals and minerals. In the long run, rising real prices and resource scarcity of these inputs are possibilities which could constrain agricultural production.

(iv) The increased use of external inputs associated with agricultural intensification can lead to pollution and the diminution of the agricultural resource base. A rising level of pollution from industrial growth can also adversely affect agricultural productivity.

(v) The ecosystem underlying certain modern agricultural practices is potentially vulnerable to shocks and stress. Irreversible environmental resource degradation could lead to the collapse of production.

Essentially, sustainability is concerned with the preservation of agricultural productive capacity for the foreseeable future. Productive capacity includes man-made capital, technological know-how and labour skills as well as land and other non-reproducible assets. The total capital stock, or value, of the system may not depend on a single combination of factors forming the productive capacity—it may be possible to preserve the total capital stock of the system using various different factor combinations, i.e. allowing substitution between the various factors of production. The potential for technological change to widen substitution possibilities is an unresolved issue in the sustainability debate. A related issue specific to agricultural sustainability concerns the extent to which sustainability in farming systems depends on technical progress in industries upstream from agriculture.

Here we do not attempt to confront broad issues but concentrate on appropriate measures of the performance of an agricultural system as indicators of sustainability. In a limited sense, the ability of an agricultural cropping system to maintain a consistent flow of marketable output over time is an indicator of sustainability. However, this indicator ignores the mix and quantities of inputs used in conjunction with land, the basic natural capital. (By land we mean the pedological characteristics of the plot in conjunction with its space/site relationship to its environment—sunlight, air and rain. As a consequence, soil fertility is likened to a capital stock capable of providing an annuity flow). It also fails to account for (a) possible environmental degradation, or both on-site and off-site negative externalities and (b) non-market benefits associated with the cropping system, or positive externalities. For operational purposes, sustainability might be assessed in relation to an index of total factor productivity (TFP) with due allowance for externalities such that 'a sustainable system has a non-negative trend in total factor social productivity over time' (Lynam and Herdt, 1989). Over the long term a sustainable system will exhibit resilience in the face of external events. Thus it will be able to maintain constant productivity (or a positive trend in productivity) by using technology that resists both cumulative stresses such as soil erosion and can withstand over time the impact of negative transitory shocks such as drought. Shocks will of course lead to short run variability in productivity, so it will be necessary to assess and quantify the trends in TFP by statistical methods. Finally, the stability of TFP (around a trend) might also be deemed to be a relevant indicator of system performance. Again statistical procedures are required to provide information on the dispersion of TFP around a trend.

Quantifying movements in total factor social productivity over time is no easy task because of information requirements. A time series database on quantities of output and inputs must be compiled; weights for combining outputs (inputs) into a single index of production (resource use) have to be determined; index number issues have to be addressed and externalities evaluated.

## 3.2 MEASURING THE PRODUCTIVE PERFORMANCE OF A SYSTEM OVER TIME

This section draws on Capalbo and Antle (1988) and Lingard and Rayner (1975). The *activity of production* may be defined as the process of combining and co-ordinating materials and factor services (inputs) in the creation of goods and services (outputs). Economists perceive this transformation process through the concept of the *production function* which relates the outputs from a production activity to the quantity of inputs combined in that activity over a given production period. For example, in monocultural wheat cropping the outputs are the joint products grain and straw, whilst in a multicropping system the outputs might be products from oil-seeds, barley and wheat. For both systems, the inputs producing outputs might be land, labour services, machinery services, seeds and crop treatments (fertiliser/pesticides).

For a single output, multiple input production activity, the production function depicts the *technology* of production in terms of two principal characteristics:

(i) the ease with which inputs may be substituted for each other in varying proportions to produce a given output; and
(ii) the relationship between output and given quantities of inputs.

A *change in the state of technology* is said to occur when there is a shift in the relationship between output and given inputs. *Disembodied* technological change refers to a situation where improved methods of utilising existing resources lead to a higher output for a given bundle of inputs. For example, a change in the timing of fertilizer application might result in improved yield for no change in inputs. Technological innovation is frequently *embodied* in design improvements in existing inputs, for example if a more efficient fertilizer with increased nutrient content is developed. Embodied technological change implies that input *quality* alters and that attention is paid to the correct *measurement* of input service flows. Quality change is a general phenomenon affecting all chemical inputs, seeds, capital and labour utilized in agricultural production. Finally, technological innovation may refer to the introduction of new production processes so that the basic form of the production function alters or to the use of new inputs so that the input set employed differs from that used under the old technology.

Since technological change can imply changes to both input productivities and utilization, a distinction is often made between (Hicks) *neutral* and *non-neutral* innovation. Neutral technical change is said to take place if the transformation of inputs into output alters without affecting substitution possibilities between inputs. Conversely, change is non-neutral if it affects substitution possibilities. Under non-neutral innovation, holding total cost constant, factor cost shares change so that the innovation may save on or increase the relative usage of a factor input.

A concept closely related to the notion of technological change is that of *productivity growth*. If technical change increases the effectiveness by which a given input combination is transformed into output, then there is said to be an increase in productivity of the bundle of inputs or aggregate input. *Productivity measurement* attempts to provide a number or score over time of output in relation to inputs used; that is, it assesses movements in output/input ratios. Productivity is said to increase when larger quantities of output are produced from the same quantities of inputs or when the same amount of output is produced with smaller quantities of inputs.

The ratio of output to a single input is called the partial productivity of that input; for example, crop yield is the partial productivity of the land input for the crop in question. However, single factor productivity is a misleading measure of gains in productive efficiency since it does not allow for factor substitution. Consequently, economists focus on *total factor productivity* (TFP). TFP is a ratio of an index of aggregate output to aggregate inputs. By constructing an index of TFP, it is possible to assess objectively the productivity performance of the system over time. The ratio of TFP between a pair of years, denoted by $r$ and $s$ is defined by:

$$\frac{TFP_s}{TFP_r} = \frac{Q_s/X_s}{Q_r/X_r} \tag{3.1}$$

or

$$\ln\left(\frac{TFP_s}{TFP_r}\right) = \ln\left(\frac{Q_s}{Q_r}\right) - \ln\left(\frac{X_s}{X_r}\right) \tag{3.2}$$

where

$Q_t$ is an index of aggregate output in year $t = r, s$;
$X_t$ is an index of aggregate input in year $t = r, s$.

The aggregator functions $Q$ and $X$ can be written as

$$Q_t = f(Q_{1t}, \cdots Q_{mt}, u_1 \cdots u_m) \quad \text{for } t = r, s$$
$$X_t = f(X_{1t}, \cdots X_{nt}, v_1 \cdots v_n) \quad \text{for } t = r, s \quad (3.3)$$

where $m$ individual outputs are produced from $n$ individual inputs and the weights $u_j$ and $v_i$ ($j = 1 \cdots m, i = \cdots n$) used in the aggregator functions are held constant between $t = r$ and $t = s$. For example, the simplest aggregator function is a weighted mean where

$$Q_t = f(Q_{1t}, \cdots Q_{mt}, u_1, \cdots u_m) = \sum_{j=1}^{m} u_j Q_{jt}. \quad (3.4)$$

If the algebraic form of the production function is known, then that form should be used for aggregation.

The relationship between TFP measurement and specific functional forms for the production function may be conveniently illustrated for a single-output/two-input linearly homogeneous production function exhibiting Hicks neutral technical change. The production function is called linearly homogeneous (or exhibits constant returns to scale) when a proportionate expansion of all inputs leads to the same proportionate expansion of output. Output at time $t$ can be written as:

$$Q_t = A(t) f(X_{1t}, X_{2t}) \quad \text{for } t = r, s \quad (3.5)$$

where $A(t)$ is a shift factor dependent only on the passage of time and reflecting technological change, and $f(X_{1t}, X_{2t})$ is a linearly homogeneous production function in inputs $X_1$ and $X_2$. Expression (3.5) can be used to form a ratio of TFP between years $s$ and $r$ with the aggregator function $f(X_{1t}, X_{2t})$ being used to form the input index. This gives:

$$\left( \frac{TFP_s}{TFP_r} \right) = \left( \frac{A_s}{A_r} \right) = \frac{Q_s / [f(X_{1s}, X_{2s})]}{Q_r / [f(X_{1r}, X_{2r})]}. \quad (3.6)$$

Construction of the index of TFP requires knowledge of the form of $f(X_1, X_2)$ in order to generate the aggregate index of the two inputs. Furthermore, assuming both a specific form of the production function *and* competitive equilibrium in production leads to a specific aggregator function expressed in terms of input prices and input quantities. Here we consider certain production functions often used by economists.

## Leontief or fixed coefficient production function

The Leontief production function assumes that the inputs are used in fixed proportions. Specifically, a unit of output requires inputs of $a_1$ units of $X_1$ and $a_2$ units of $X_2$:

$$f(X_{1t}, X_{2t}) = \min \left( \frac{X_{1t}}{a_1}, \frac{X_{2t}}{a_2} \right); \quad a_1 > 0, \quad a_2 > 0 \quad \text{and} \quad t = r, s. \quad (3.7)$$

The minimum input requirements $X_{1t}^*$ and $X_{2t}^*$ for output $Q_t$ are

$$X_{1t}^* = \frac{Q_t}{A_t}a_1, \quad X_{2t}^* = \frac{Q_t}{A_t}a_2 \quad \text{so} \quad X_{1t}^* = \frac{a_1}{a_2}X_{2t}^* \quad \text{for } t = r, s \qquad (3.8)$$

then the minimum cost of production for output $Q_t$ is

$$C_t^* = w_1 X_{1t}^* + w_2 X_{2t}^* \quad \text{for } t = r, s \qquad (3.9)$$

where $w_1$ and $w_2$ are unit input prices assumed constant for $t = r, s$. Then

$$\frac{Q_s}{Q_r} = \frac{A_s}{A_r}\frac{X_{is}^*}{X_{ir}^*} \quad \text{and} \quad \frac{C_s^*}{C_r^*} = \frac{X_{is}^*}{X_{ir}^*} \quad \text{for } i = 1, 2$$

so

$$\frac{Q_s/C_s^*}{Q_r/C_r^*} = \frac{A_s}{A_r} \qquad (3.10)$$

and the appropriate aggregator function for inputs is the cost function (3.9) which is *linear* in prices and quantities, so

$$X_t = w_1 X_{1t} + w_2 X_{2t}. \qquad (3.11)$$

If prices change between period $r$ and period $s$, then the aggregator function is based on prices in one of these two periods.

## Cobb–Douglas or geometric production function

The Cobb–Douglas (C–D) production function allows for substitutability between inputs and is thus less restrictive than the fixed proportions functions. The C–D function is linear in logarithms:

$$\ln f(X_{1t}, X_{2t}) = a_1 \ln X_{1t} + a_2 \ln X_{2t} \qquad (3.12)$$

or

$$f(X_{1t}, X_{2t}) = X_{1t}^{a_1} X_{2t}^{a_2}$$

where the $a_i$ are known as production elasticities ($\partial \ln f / \partial \ln X_{it}$). Under competitive equilibrium, marginal products ($\partial Q_t / \partial X_{it}$) are equated to input prices:

$$p\frac{\partial Q_t}{\partial X_{it}} = pa_i \frac{Q_t^*}{X_{it}^*} = w_i \quad \text{for } i = 1, 2 \quad \text{and } t = r, s \qquad (3.13)$$

where $p$ is the price of unit output and $w_i$ is the unit input price ($i = 1, 2$) and all prices are assumed constant for $t = r, s$. $Q_t^*$ and $X_{it}^*$ ($i = 1, 2$) are the optimal quantities for $t = r, s$. If the production function is linear homogeneous, then $a_1 + a_2 = 1$ and the value of output is equal to the cost of production. Then, from (3.13),

$$a_i = \frac{w_i X_{it}^*}{pQ_t^*} = \frac{w_i X_{it}^*}{\Sigma w_i X_{it}} = s_i^* \qquad (3.14)$$

where $s_i^*$ is the optimal factor cost share. The TFP ratio may be expressed as

$$\ln\left(\frac{A_s}{A_r}\right) = \ln\left(\frac{Q_s}{Q_r}\right) - \left(s_1^* \ln \frac{X_{1s}^*}{X_{1r}^*} + s_2^* \ln \frac{X_{2s}^*}{X_{2r}^*}\right) \qquad (3.15)$$

so that the appropriate aggregator function is

$$\ln X_t = s_1 \ln X_{1t} + s_2 \ln X_{2t} \qquad (3.16)$$

Whilst expression (3.16) specifies constant factor shares, in practice a choice would have to be made between estimated factor shares in period $r$ or in period $s$.

## Translogarithmic production function

The linear (in logarithms) functional form of the Cobb–Douglas production function is relatively inflexible and from the viewpoint of production theory a quadratic functional form is more appropriate. Extending the Cobb–Douglas in this way gives the translog production function:

$$\ln f(X_{1t}, X_{2t}) = \beta_0 + \sum_{i=1}^{2} \beta_i \ln X_{it} + \frac{1}{2} \sum_{i=1}^{2} \sum_{k=1}^{2} \beta_{ik} \ln X_{it} \ln X_{kt} \qquad (3.17)$$

with the following restrictions imposed to give linear homogeneity

$$\beta_1 + \beta_2 = 1; \quad \beta_{12} = \beta_{21} \quad \text{and} \quad \sum_{k=1}^{2} \beta_{ik} = 0 \quad \text{for} \quad i = 1, 2.$$

It may be shown (Diewert, 1976) that

$$\ln f(X_{1s}, X_{2s}) - \ln f(X_{1r}, X_{2r}) = \frac{1}{2} \sum_{i=1}^{2} \left(\frac{\partial \ln f(X_{1s}, X_{2s})}{\partial \ln X_i} + \frac{\partial \ln f(X_{1r}, X_{2r})}{\partial \ln X_i}\right)$$

$$(\ln X_{is} - \ln X_{ir}).$$

Production elasticities are defined by $\partial \ln Q_t / \partial \ln X_{it}$ and, under competitive equilibrium, relative production elasticities are equal to relative input cost shares ($s_i$). Consequently, the aggregator function is written as:

$$\ln X_t = \frac{1}{2} \sum_i^2 (s_{ir} + s_{is}) \ln (X_{it}) \quad \text{for} \quad t = r, s \qquad (3.18)$$

so that factor shares are averaged over periods $r$ and $s$. The index of aggregate input given by (3.18) is known as the Tornqvist–Theil index (see below).

The above examples illustrate the correspondence between specific functional forms of the production function and the appropriate aggregator function for inputs in measuring TFP under the assumption of competitive equilibrium. Since it would be unusual to have exact knowledge of the production function, economists have usually advocated measurement by an aggregator functional form that corresponds to a flexible functional form for the production function. In addition, the assumption of competitive behaviour is required to derive results. In order to avoid these problems with the production function approach to TFP measurement, an alternative is to rely directly on an index number approach.

## 3.3 INDEX NUMBERS

Consider a very simple production process where a single output ($q$) is produced from a single input ($x$). Then productivity growth between periods $r$ and $s$ is

$$\frac{TFP_s}{TFP_r} = \frac{q_s/x_s}{q_r/x_r}.$$

The output growth rate between $r$ and $s$ is $q_s/q_r$ and the input growth rate is $x_s/x_r$. If output exceeds input growth, then there is a productivity improvement over the time period $r$ to $s$.

In the multiple ouput, multiple input case, the output ratio is replaced by an output quantity index $Q_s/Q_r$ and the input ratio by the input quantity index $X_s/X_r$ where $Q_s$ and $Q_r$ ($X_s$ and $X_r$) are index numbers aggregated over individual outputs (inputs).

Construction of the aggregate output (input) index requires a choice of (a) the weights and (b) the specific functional form for the aggregator function. Relevant prices are employed to define the weights in the economic approach to productivity measurement since monetary value is a natural common unit of output (input) measurement for the purposes of aggregation. Using monetary value as the measuring rod is 'natural' since the ratio of any two output prices represents relative marginal valuations placed on the products and the ratio of any two input prices represents relative marginal costs of the inputs in a non-distorted competitive economy. Thus prices should provide information about the relative 'worth' of goods and factor services.

But in the usual situation where relative output (input) prices change between the time periods $r$ and $s$, different indexes may be obtained, depending on the base period chosen for defining the constant price weights (or weights derived from prices). For example, either prices in period $r$ or prices in period $s$ or some average of prices in the two periods might be chosen as the weights. Similarly, alternative functional forms can generate different implied productivity outcomes. The arithmetic and the logarithmic functional forms are the most widely employed in practice.

Two of the most commonly used arithmetic indexes are the Laspeyres and the Paasche indexes. Assuming that $r$ is the base period and $r < s$, then the Laspeyres index uses prices in period $r$ as weights:

$$\text{Laspeyres}\left(\frac{Q_s}{Q_r}\right) = \frac{\sum_{j=1}^{m} P_{jr}Q_{js}}{\sum_{j=1}^{m} P_{jr}Q_{jr}}; \quad \text{Laspeyres}\left(\frac{X_s}{X_r}\right) = \frac{\sum_{i=1}^{n} W_{ir}X_{is}}{\sum_{i=1}^{n} W_{ir}X_{ir}}$$

where $P_{jr}$ and $W_{ir}$ are the prices of the $j$th output and the $i$th input respectively in period $r$ (the base period).

The Paasche index uses prices in period $s$ as weights:

$$\text{Paasche}\left(\frac{Q_s}{Q_r}\right) = \frac{\sum_{j=1}^{m} P_{js}Q_{js}}{\sum_{j=1}^{m} P_{js}Q_{jr}} \quad \text{Paasche}\left(\frac{X_s}{X_r}\right) = \frac{\sum W_{is}X_{is}}{\sum W_{is}X_{ir}}$$

where $P_{js}$ and $W_{is}$ are the prices of the $j$th output and the $i$th input respectively in period $s$ (the 'current' period).

Given that prices will not move proportionately between periods $r$ and $s$, various averages of the Laspeyres and Paasche indexes have been proposed. The most widely used is the Fisher index, defined as the geometric mean of the Laspeyres and Paasche indexes. The Fisher aggregate output index is defined as:

$$\text{Fisher}\left(\frac{Q_s}{Q_r}\right) = \left[\left(\frac{\sum P_{jr}Q_{js}}{\sum P_{jr}Q_{jr}}\right)\left(\frac{\sum P_{js}Q_{js}}{\sum P_{js}Q_{jr}}\right)\right]^{1/2}$$

with a similar expression derived for the Fisher aggregate input index.

Finally, the most widely used logarithmic index is probably the Tornqvist–Theil or translog index. The Tornqvist aggregate quantity index is defined as:

$$\text{Tornqvist } \ln\left(\frac{Q_s}{Q_r}\right) = \frac{1}{2}\sum_{j=1}^{m}\left(\frac{P_{jr}Q_{jr}}{\sum P_{jr}Q_{jr}} + \frac{P_{js}Q_{js}}{\sum P_{js}Q_{js}}\right)\ln\left(\frac{Q_{js}}{Q_{jr}}\right) = \frac{1}{2}\sum_{j=1}^{m}(s_{jr} + s_{js})\ln\left(\frac{Q_{js}}{Q_{jr}}\right)$$

where $s_{jr}$ and $s_{js}$ are output revenue shares. A similar expression is defined for the Tornqvist aggregate input index, with the $s_{jr}$ and $s_{js}$ representing input factor shares.

The choice of a particular index is often justified with respect to statistical and economic properties. The statistical approach assesses biases of a particular index in relation to stipulated tests: the Fisher index is usually superior in this respect to the three other indexes defined above (Diewert 1989). As we noted above, the economic approach interprets productivity growth as a shift in the underlying production function, and the Tornqvist index based on the quadratic logarithmic functional form is superior to indexes based on the linear (arithmetic or logarithmic) functional form since it is consistent with a flexible production function which does not arbitrarily constrain the substitution possibility between inputs. Nevertheless, over a short time period, a simple index such as the

Laspeyres can often provide an acceptable (first order) approximation to a flexible aggregator function. Indeed for agronomic trials where inputs are frequently used in roughly fixed proportions, at least for relatively short periods of time, the Laspeyres index may well be a satisfactory representation of the underlying production function (see Section 3.2), particularly when used in conjunction with chaining.

The discussion of index numbers has so far been confined to comparisons between two time periods only. For productivity measurement over a long period of time, *chaining* indexes for successive time periods, or at least changing the weights frequently, is recommended. With chain-linking, an index is calculated for each two successive periods $t$ and $t + 1$, over the whole sample from time $t = 0$ to $t = T$ and the separate indexes are then multiplied together:

$$I(0, T) = I(0, 1).I(1, 2) \ldots I(t, t + 1) \ldots I(T - 1, T).$$

Chain-linking means that the index takes account of changes in relative values/costs throughout the period of interest. This procedure has the advantage that no single period plays a dominant role in determining price or share weights and biases are likely to be reduced. So, for example the Laspeyres index with frequently revised weights can often given similar results to the Tornqvist index.

When a new output (input) is introduced into the production process the linking of two indexes overlapping the period of introduction is required. Thus an index is calculated without the new factor prior to and for a short period covering adoption and an index is calculated from the introduction of the factor to a period after adoption. The two indexes are then linked together from data on the overlap.

Constructing aggregate output and input indexes of whatever form in order to assess the long-run trend in TFP requires an adequate database on outputs produced from and inputs used in the agricultural system. In the present context, probably the two most difficult issues to resolve are making allowance for quality changes in outputs and inputs and estimating the service flows from capital inputs, especially machinery.

## 3.4 MEASURING PROFITABILITY

The previous sections of this chapter consider methods of measuring productivity maintenance over time, using price weights either fixed from a base year or changed via chain-linking. In either case, the indexes examine patterns over time on some relative scale. In reality, although it is important that productivity is maintained in the long term, in the short term an agricultural system must at least meet its costs in order to survive. For that reason, the *profitability* (ratio of aggregated output values to input costs) of a system must also be considered. In this case, we define profitability in year $t$ to be simply

$$\Pi_t = \frac{\sum Q_{jt} P_{jt}}{\sum X_{it} W_{it}} = \frac{R_t}{C_t}$$

where

$Q_{jt}$ is the quantity of output $j$ produced at time $t$ with unit value $P_{jt}$, and $R_t$ is total value of output;
$X_{it}$ is the quantity of input factor $i$ used at time $t$ with unit cost $W_{it}$, and $C_t$ is total cost.

We use a ratio rather than a simple difference in order to make the measure independent of scale. Since, in competitive equilibrium, the profitability of a system would be equal to unity by definition, we do not include the cost of buying or renting land as an input cost. Consequently, the residual difference between the value of output and input cost is interpreted as rent to land, with land rent $r_t = R_t - C_t$. Consequently, the profitability ratio is:

$$\Pi_t = 1 + \frac{r_t}{C_t}.$$

## 3.5 TREND AND VARIABILITY IN TFP AND SUSTAINABILITY IN AN AGRONOMIC TRIAL

Constructing an index of TFP for a long-term agronomic trial permits the evaluation of the ability of the cropping system to maintain productivity over time. In conjunction with a measure of profitability, i.e. the short-term viability of the system, we define sustainability as follows: an agricultural system is sustainable if it can maintain productivity over time (non-negative trend in the TFP index) and if the value of output generally exceeds the value of variable and quasi-fixed inputs (profitability not smaller than unity). Both measures are required to avoid classification of a non-profitable system which maintains productivity as sustainable. The index is set at 100 for a particular year—the base year—of the trial and the output to input ratio is computed for all other years relative for the base. Recall from Section 3.2 that the ratio of TFP between any pair of years, denoted by $r$ and $s$ is defined by

$$\frac{TFP_s}{TFP_r} = \frac{Q_s/X_s}{Q_r/X_r}. \tag{3.19}$$

If year $r$ is taken as the base year and the numerical value of the rato $(TFP_s/TFP_r)$ is equal to $\theta$ then $TFP_s = 100\,\theta$.

If TFP for year $s$ is less (more) than 100, then productivity has fallen (risen) between years $r$ and $s$; that is, less (more) output is being produced from a given quantity of inputs in year $s$ compared with the base year.

In this manner, a TFP index is computed for the sequence of $T$ years of the trial relative to the base year. Year to year fluctuations in TFP will result from natural variations in yield, largely weather induced. Overall, a sustainable cropping system will exhibit mean TFP rising, or at least not falling, over the trial period of $T$ years. Figure 3.1 depicts four possible paths for TFP, with year 0 as the base year and with short term variation removed. Path A shows mean TFP rising monotonically implying clear sustainability. Path B depicts an initial fall in TFP and then a rising trend and again implies sustainability. Conversely, Path C and D depict non-sustainability.

Figure 3.2 depicts the possible response of TFP to severe stress, such as a prolonged drought (again removing short term variability). Path A shows recovery of TFP from stress indicating a resilient system whilst Path B indicates irreversible degradation of the resource base and a non-sustainable system.

Finally, Figure 3.3 incorporates natural variability and contrasts the degree of stability in two possible time paths: Path A is highly stable whilst Path B displays considerable instability.

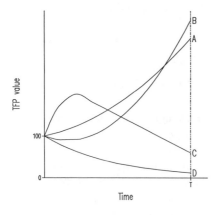

**Figure 3.1** Possible time paths for TFP: trends

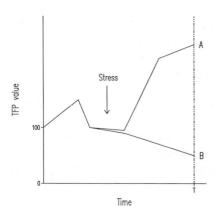

**Figure 3.2** Possible time paths for TFP: resilience

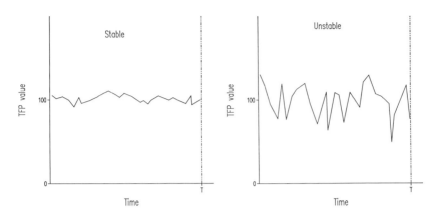

**Figure 3.3** Possible time path for TFP: stability

Quantifying the time path for the mean of TFP and assessing variability around a changing mean requires the fitting of a trend to the TFP index by statistical measures. This involves issues such as the appropriate form of the trend function, the detection and treatment of possible outliers, the significance of the estimated parameters and the characterisation of instability in terms of dispersion and skewness of residuals from the trend.

## 3.6 STATISTICAL ASSESSMENT OF TFP INDEXES

For any system under consistent management, the TFP index will consist of an underlying trend plus natural year-to-year variability, reflecting changes in yields due to changing weather conditions, as in Figure 3.3. Given a TFP series, statistical methods can be used to estimate both the trend and the variability of the series.

It is not necessarily the overall trend of an index series that may be of interest. A series may in fact consist of several subseries which each have different trends. The subseries may correspond to changes in management practice or changes in technology, in which case a change in trend would be expected. These subseries should be treated separately in order to get unbiased estimates of trend for individual subseries. Alternatively, the subseries may reflect an underlying change in the resource base, for example where a system has 'mined' some natural resource to the point of exhaustion and productivity can no longer be maintained. In this case, the change in trend would be unexpected and possibly downwards and the aim of the analysis would be to detect any significant change in trend as soon as possible. This gives two statistical problems to be solved: estimation of trend in identifiable subseries; and detection of unexpected changes in trend in an ongoing series.

### 3.6.1 Estimation of index trend

For an index series over time relating to a system under reasonably stable conditions, i.e. no large changes in management practice or technology, the trend must be quantified in order to classify the system as sustainable or unsustainable. Since there is no theoretical underlying form that the index series should take, the estimation process must rely on empirical modelling. One approach (Barnett et al., 1993) uses an additive model (Hastie and Tibshirani, 1990) to fit the average linear slope over the series as an estimate of trend, plus a smooth non-parametric function to describe and take account of the non-linear trend component. If the index at time $t$ takes value $y_t$, the fitted model has the form

$$y_t = a + bt + f(t) + e_t$$

where $a$ is the intercept and $b$ the slope of the average linear trend of the series, $f(t)$ represents the non-linear component of the trend and $e_t$ represents random error. It is assumed that the errors $e_t$ are independent normal random variables with equal variance. If there is evidence of changing variation over time, then some suitable monotonic transformation should be used to stabilize the variance before analysis. After estimation, the residuals are used to generate an estimate of the error variance for the subseries considered. This estimate of error is a measure of stability for the series: a small error variance indicates a relatively stable system. However, note that comparisons of variation between systems can only be made if the same transformation or no transformation has been applied to each series considered. The additive model gives a more appropriate measure of variation and stability than a purely linear model, since the residuals obtained from fitting a linear trend alone contain a 'lack-of-fit' component due to any non-linear systematic trend. If this non-linear trend component is significant, the error variance is overestimated by the linear model. This also means that the statistical significance of the average linear trend will be decreased. ignoring the non-linear trend also reduces the chance of detecting any linear trend, if it is present. If the non-linear systematic component is not significant, the model reduces to simple linear regression. Given the approximate degrees of freedom used in fitting the smooth function $f(t)$, smay, then the test statistic for testing the null hypothesis $b \geqslant 0$ is $b/ese(b)$, where $ese$ is the estimated standard error. Under the null hypothesis, the test statistic has, approximately, a $t$-distribution on $T - 2 - s$ degrees of freedom.

### 3.6.2 Detecting a change in trend

Given an index series for an ongoing agricultural system it is desirable to monitor the system for signs of change, especially changes towards a downwards trend in the index series. One simple base against which trend can be monitored is the extrapolation of any significant current linear trend in the series. The behaviour of the non-linear component of trend cannot be used in this context since it is non-parametric and cannot be extrapolated into the future for comparison with actual data values. If there is no evidence of significant linear trend, future values should be compared with the average series value. Confidence intervals around the predicted future values can be calculated using estimates of error variation from the model and a set of future index values outside these limits is evidence of a change in trend. Stability of an index series is important in this context, since if large deviations from the linear trend exist, the confidence intervals may become very large and tests will be insensitive to changes in trend.

Where the index is stable, with no evidence of trend, diagnostic plots such as CUSUM (Johnson and Leone, 1976), using the mean as the target value, can be used to detect a shift in the mean value of the series empirically and hence a possible change in trend. Where a linear trend has been found, a CUSUM plot based on deviations from fitted values may give similar information.

The methods above illustrate how the trend and variability of an index series might be quantified. However, these methods make assumptions which may not generally hold. For example, successive values of index series are assumed to be independent. Analysis of several Rothamsted datasets (see Chapter 9 and Barnett *et al.*, 1994) suggests that there is evidence of dependence between index values only where treatments have varied systematically between years. More-over, in these cases it can be argued that the treatment effects should be spread over successive years, thus removing the pattern.

Statistical assessment of index series will be more reliable for a stable series than an unstable series: it is possible that the estimated downwards trend in a stable series (small natural variation) would be statistically significant, whereas the same downwards trend in an unstable series (large variation about the trend) might not be statistically significant, since the variability in the series masks the trend. This further illustrates the point that stability is a desirable attribute of a system subject to monitoring, since it allows accurate measurement of the trend and early detection of any changes in trend. Stability also implies that a system can withstand transitory shocks and that the risk of large losses in any one one year (albeit offset by large gains in another) is reduced. This will be especially important in situations where the farmer (or society) does not have the resources to tide over a bad year.

## 3.7 CONCLUSIONS

If information is available about the form of the production function underlying an agricultural system, then the form of the TFP to be calculated can be deduced. Usually, this function is unknown and an index number approach must be taken. From a theoretical standpoint a Tornqvist–Theil index, which corresponds to a flexible (although unknown) underlying production function, is the most suitable. However, for agronomic trials where factor proportions may be relatively unchanged, at least for short periods of time, a Laspeyres index with frequent change of price weights through chain-linking may well be satisfactory.

Sustainability can be defined in terms of several simple components: the TFP index series must have a non-negative underlying trend and the system must be profitable. Stability—relatively small variation around the trend—is another desirable feature since it allows accurate determination of the trend and gives improved efficiency in detecting changes in trend. Within this chapter, the TFP has been defined in terms of input and output factors. Given objective and accurate quantification, some of these factors could represent externality effects, leading to a measure of TSFP (Total Social Factor Productivity). However, quantification of these effects is at present ill defined and must be improved realistic TSFP indexes can be produced.

# PART II
*Detailed Studies*

# 4
# Long-term Cotton Productivity Under Organic, Chemical, and No Nitrogen Fertilizer Treatments, 1896 to 1992

G. Traxler, J. Novak, C. C. Mitchell, Jr., and M. Runge

*Auburn University, Alabama, USA*

The Old Rotation experiment at Auburn University, Alabama (1896) is the oldest continuous cotton experiment in the United States (and possibly the world), the third longest-running continuous field crop experiment in the United States, and the first experiment to demonstrate the benefits of rotating cotton with other crops and to measure the contributions of nitrogen-restoring winter legumes to a cotton production system. The experiment continues to document the long-term effect of these cotton production systems. Because only minor changes have been made in the cropping systems, the record of fertilizer use and yields provides information on the fundamental problem of maintaining soil fertility and sustaining crop production in soils of the Southeastern United States.

---

*Agricultural Sustainability: Economic, Environmental and Statistical Considerations.*
Edited by V. Barnett, R. Payne and R. Steiner © 1995 John Wiley & Sons Ltd

## 4.1 HISTORIC PERSPECTIVE ON 'THE OLD ROTATION'

In February, 1883, the Alabama legislature enacted the enabling legislation for what is currently known as the 'Old Rotation'. The trustees of the Agricultural and Mechanical College of Alabama (now Auburn University) were directed to '... establish and maintain an agricultural farm or station where careful experiments shall be made in scientific agriculture.' To fulfil this mandate the college trustees purchased 91.53 hectares of land for farm research plots, part of which is the site of the Old Rotation.

The agricultural economy of the Southeast was heavily dependent on cotton when the Old Rotation was established in 1896. At that time, more than 9.3 million hectares of cotton were planted across the South, with a regional average lint yield of less than 224 kg/ha (US Dept. of Agriculture, 1899). Auburn's Professor John F. Duggar established the Old Rotation to test his belief that Alabama soils and climate could sustain more profitable yields of cotton with minimum fertilization if a reasonable rotation with legumes could be worked out. Today the Old Rotation is maintained by the Department of Agronomy and Soils and the Alabama Agricultural Experiment Station as a continuing test of Duggar's hypothesis.

## 4.2 OLD ROTATION TREATMENTS

Three of thirteen Old Rotation treatments (Table 4.1, Figure 4.1) were analysed in this study in order to compare the effect of the use of nitrogen fertilizers and of crop rotation on long-term productivity. The treatments were:

1. continuous cotton with no nitrogen and no winter legume (plot 6 in Table 4.1; referred to hereafter as the no N treatment)
2. continuous cotton with a winter legume (crimson clover and/or vetch) used as a green manure (plot 8; the winter legume treatment), and
3. continuous cotton with annual application of 134 kg/ha of nitrogen (plot 13; the 134 kg/ha of N treatment).

The first two treatments have not changed since the experiment was initiated in 1896. The third treatment was in a two-year cotton–vetch–cowpea rotation until 1955. Analysis of these three treatments allows the long-term performance of systems using a legume rotation to be contrasted with that of a system relying exclusively on chemical sources of nitrogen, and with a system without benefit of either organic or chemical nitrogen sources.

## 4.3 SITE CHARACTERISTICS AND SOIL MEASUREMENTS

As in most experiments conducted at the turn of the century, treatments were not replicated. The original design of the study included 13 non-replicated treatments

Site characteristics and soil measurements 43

**Table 4.1** Crops and fertilizer rates (kg/ha N-P-K) used in 'Old Rotation' since 1896

| Treatment/plot | 1896–1924 | 1925–1931 | 1932–1947 | 1948–1955 | 1956–present |
|---|---|---|---|---|---|
| 1 | corn 0-11-18<br>cowpeas | corn 0-13-18<br>vetch 0-31-0 | cotton 0-36-56<br>vetch | cotton 0-36-56 | cotton 0-40-56 |
| 2 | corn 0-11-18 | corn 0-44-18 | cotton 0-36-56 | cotton 0-36-56<br>vetch | cotton<br>vetch/clover 0-40-56 |
| 3 | cotton 0-11-18<br>vetch | cotton 0-13-18<br>vetch 0-31-0 | cotton 0-18-28<br>vetch 0-18-28 | cotton 0-18-28<br>vetch 0-18-28 | cotton 0-20-28<br>vetch/clover 0-20-28 |
| 4, 7 | cotton 0-11-18<br>vetch<br>corn 0-11-18<br>cowpeas | cotton 0-13-18<br>vetch 0-31-0<br>corn 0-13-18<br>vetch 0-31-0 | cotton 0-18-28<br>vetch 0-18-28<br>corn 0-18-28<br>vetch 0-18-28 | cotton 0-18-28<br>vetch 0-18-28<br>corn 0-18-28<br>vetch 0-18-28 | cotton 0-40-28<br>vetch/clover 0-20-28<br>corn 0-0-0<br>vetch/clover 0-20-28 |
| 5, 9 | cotton 0-11-18<br>vetch<br>cowpeas 0-11-18 | cotton 9-13-18<br>vetch 0-31-0<br>cowpeas hay 0-13-18<br>vetch 0-31-01 | cotton 0-18-28<br>vetch 0-18-28<br>cowpea hay 0-18-28<br>vetch 0-18-28 | cotton 0-18-28<br>vetch/clover 0-18-28<br>cowpea hay-18-28<br>vetch 0-18-28 | cotton 134-40-56<br>vetch/clover 0-20-28<br>corn 134-0-0<br>vetch/clover 0-20-28 |
| 6 | cotton 0-11-18 | cotton 0-44-18 | cotton 0-36-56 | cotton 0-36-56 | cotton 0-40-56 |
| 8 | (same as #3) | (same as #3) | cotton 0-36-56<br>vetch | cotton 0-36-56 vetch | cotton 0-40-56<br>vetch/clover |
| 10, 11, 12 | cotton 0-11-18<br>vetch<br>corn 0-11-18<br>cowpeas/oats<br>cowpeas 0-11-18 | cotton 0-44-18<br>vetch<br>corn 0-44-18<br>oats<br>cowpea hay 0-44-18<br>vetch | cotton 0-18-28<br>vetch 0-18-28<br>corn 0-18-28<br>oats 0-18-28<br>cowpea hay 0-18-28<br>vetch 0-18-28 | cotton 0-18-28<br>vetch 0-18-28<br>corn 0-18-28<br>oats 0-18-28<br>cowpea hay | cotton 0-40-56<br>vetch/clover 0-40-56<br>corn<br>rye 67-0-0<br>soybeans |
| 13 | (same as #5) | (same as #5) | (same as #5) | cotton 0-18-9<br>vetch 0-18-9<br>cowpea hay 0-18-9<br>vetch 0-18-9 | cotton 134-40-56 |

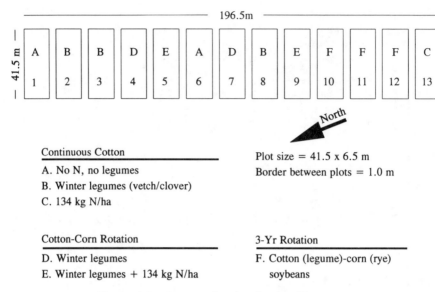

**Figure 4.1** Current plot plan for the old rotation

in plots 6.5 × 41.4 metres with a 1 metre alley. Modification of the original treatments has resulted in some treatment replication today (Table 4.1, Figure 4.1).

The Old Rotation is located at the juncture of the southern Piedmont Plateau and the Gulf Coastal Plain physiographic regions in east-central Alabama (85°W, 32°N). Average annual precipitation at the site is approximately 134.6 centimetres with a mean annual temperature of 18 degrees Celsius and 221 frost-free days.

Soils in this location are sandy Coastal Plain sediments overlying finer textured, highly weathered Piedmont soils. Although the soil at the Old Rotation site is currently identified as a Pacolet sandy clay loam (clayey, kaolinitic, thermic Typic Hapludults), it has been called a Norfolk fine sandy loam (fine-loamy, siliceous, thermic Typic Kandiudults). The site appears on the local soil survey as a Marvyn loamy sand (fine-loamy, siliceous, thermic Typic Kanhapludults), a Coastal Plain soil. Confusion arises because the site is on a gradual slope (1–3%) and the soil texture changes. The upper part of the site (plot 1) is more characteristic of the Marvyn soil (Coastal Plain) and the lower part (plot 13) is more characteristic of the Pacolet (Piedmont).

Because the Old Rotation experiment is primarily a crop rotation and legume nitrogen study, annual rates of phosphorus (P) and potassium (K) applied to each plot have been held constant across treatments. Phosphorus was applied as ordinary superphosphate (18–20% $P_2O_5$) in the early years of the experiment and as concentrated superphosphate (45% $P_2O_5$) since. Potassium was applied as kainit until 1943 and as muriate of potash (60% $K_2O$) since. Ammonium nitrate (33.5% N) has been applied on the 134 kg/ha N treatment since 1956. Nutrients were applied to the summer crop, the winter legumes, or split between the two. Where no legume or nitrogen fertilizer was added to the soil, continuous cotton production removed about 14.56 kilograms of nitrogen per hectare per year in lint and seed. This was equal to the amount of nitrogen available from atmospheric and soil microbial fixation.

Davis (1949) discussed the crop growth problems encountered on the Old Rotation during its first 50 years. Most of these, particularly K deficiencies, resulted from low applications of K fertilizers and removal of cowpea hay. Phosphorus deficiencies in the winter legume often led to low biomass production and low yielding cotton (due to less N from the legume). This observation led to split P and K fertilizer applications, which continues today. Both soil P and K have risen to high levels on these plots, thus these deficiencies are no longer observed (Figures 4.2 and 4.3).

Using data from the Old Rotation, Davis (1949) noted that '... cotton as a crop does not deplete the soil or run it down excessively. The cultural practices of leaving the land bare through the winter and of not preventing erosion are responsible for the generally low fertility level of many soils on which cotton is grown.' No records of soil measurements have been found before 1950. However, since then, periodic plough-layer soil samples have been taken for pH and Mehlich-1 (dilute double acid) extractable P and K. Soil pH values are presented in Figure 4.4 for the three treatments. There are statistical differences (p < 01) in pH and extractable P among the three treatments but no difference in extractable K.

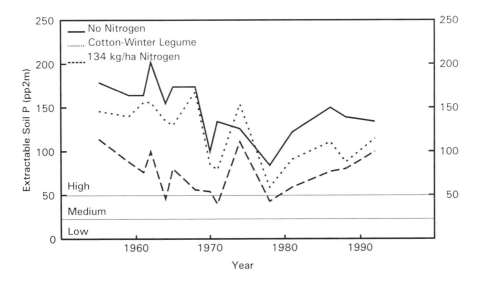

**Figure 4.2**  Soil phosphorus

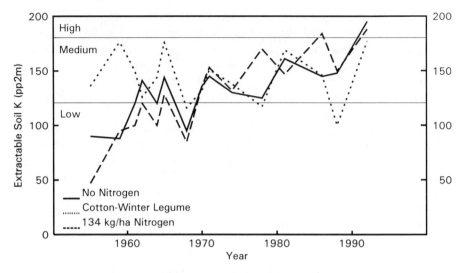

**Figure 4.3** Soil potassium

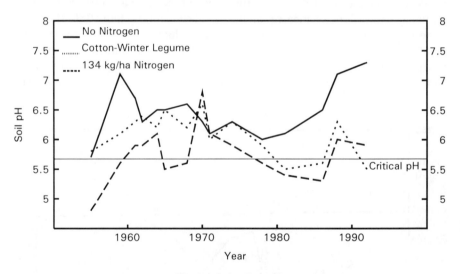

**Figure 4.4** Soil pH

The 134 kg/ha N plot is the last plot in the experiment and is at the lowest elevation. The surface soil texture tends to be somewhat finer than that of the first few plots of the experiment which are more typical of Coastal Plain soils. The apparent lower P may result from the higher P fixing capacity of the finer-textured soil in this plot.

**Table 4.2** Plough-layer soil organic matter found in selected treatments of the Old Rotation

| Treatment | Year 1988 | 1992 |
|---|---|---|
| | % organic matter* | |
| Continuous cotton | | |
| – No N, no legume | 0.7 | 0.8 |
| – Winter legume | 1.7 | 1.5 |
| – 134 kg/ha N | 1.4 | 1.8 |

*organic matter $\times\,0.53$ = organic carbon

Since 1950, individual plots have been limed (using finely ground, dolomitic limestone) whenever the soil pH dropped below 5.8, the critical pH used in Alabama for loamy soils. In spite of this effort, the 134 kg/ha N treatment (33.5% N) has tended to have a more acid reaction than either the no N treatment or the winter legume treatment.

Although the Old Rotation involves organic N sources (winter legumes), soil organic matter levels were not recorded until 1988 (Table 4.2). These recent analyses indicate important differences in long-term accumulations of stable organic carbon, which may account for some of the changes in soil aggregate structure in certain plots.

Although soil structure changes have not been quantified, agronomists have observed a noticeable decline in surface soil aggregate structure in the 134 kg/ha N treatment. Surface soil crusting following planting has been more severe on this plot than any of the others. In 1986, no cotton was harvested from this plot because of a poor stand resulting from severe surface crusting. Declining soil organic matter since 1956, when the plot was converted from a cotton–vetch–cowpea–vetch rotation to continuous cotton, may be a factor in this apparent surface crusting. On the other hand, crusting has not been as severe in the no N treatment in spite of the fact that this plot has the lowest organic matter content.

## 4.4 PRODUCTIVITY ANALYSIS

The goals and methods for the analysis of the Old Rotation are similar to those of the other chapters of this volume. Data from three of the Old Rotation plots were used to construct output, input, and productivity indexes (see Chapter 3). Trends in the index series were analysed for the length of the respective treatments (1896–1991 for the no N and winter legume treatments, 1956–1991 for the 134 kg/ha N treatment).

The advantage of using indexes is in the relative ease with which performance measures can be developed and compared. The primary interest in addressing the question of the sustainability of cotton production is in the effect on the movement of a total factor productivity index over time. Total factor productivity (TFP) is a more informative indicator than partial productivity measures, such as output per unit of land or output per unit of labour. The intuitive appeal of the TFP index is that it can be interpreted as 'output per unit of input'. As constituted here, this index number formula adjusts for the effect of changing input prices, so that any change in TFP can be attributed to a change in production efficiency, rather than changing market prices; a doubling of TFP implies that twice as much output is derived from each 'unit' of input.

The Tornquist approximation to the Divisia index (referred to hereafter as the Divisia index) was used in our analysis. Divisia input, output and total factor productivity (TFP) indexes can be calculated as:

$$I(X)_t = \prod_i (X_{it}/X_{i,t-1} 1)^{(S_{it} + S_{i,t-1})/2} \quad (4.1)$$

$$I(Y)_t = \prod_j (Y_{jt}/Y_{j,t-1})^{(W_{jt} + W_{j,t-1})/2} \quad (4.2)$$

$$TFP_t = I(Y)_t/I(X)_t \quad (4.3)$$

where

$I(X)_t$ and $I(Y)_t$ are quantity indexes of input and output use in time $t$;
$X_{it}$ is quantity of input $i$;
$Y_{jt}$ is the quantity of output $j$;
$S_{it} = r_{it}X_{it}/\sum r_{it}X_{it}$ where $r_{it}$ is the price of input $i$, and
$W_{jt} = p_{jt}Y_{jt}/\sum p_{jt}Y_{jt}$ where $p_{jt}$ is the price of output $j$.

The Divisia index number formulation has appealing theoretical properties, including consistency with an assumed underlying homogeneous translog production technology (Christensen, 1975). Although Christensen warns that the Divisia and Laspeyres indexes may diverge when the period of analysis is very long, or when 'large' changes occur in either prices or quantities, several studies have found the difference to be small (Boyle, 1988; Ball, 1985; Sidhu and Byerlee, 1992). This study used a two-year chain-linked divisia index formula, with 1990 as the base year.

The use of index numbers to analyse productivity changes of a single plot of land is unique to this project. Most previous applications have used aggregate data to analyse agricultural performance at the state, national or multinational level (Boyle, 1988; Ball, 1985; Jorgenson, et al., 1987; Lu, et al., 1979). A few studies (Sidhu and Byerlee, 1948; Cooke and Sundquist, 1989, 1991) have used survey data to analyse the performance of 'average', or representative producers.

## 4.5 OLD ROTATION DATA

The original 'Old Rotation' records for the years 1896–1919 were destroyed when a fire razed Comer Agricultural Hall in 1920. Data for 1896–1915 had been published and were recovered; data from 1916–19 are still missing. Other gaps that exist in the yield data are 1925, due to drought, 1967, and 1972–77. In this analysis, most of these gaps in the data are filled in using forecast yields obtained by regressing Old Rotation data on yields from the nearby Cullars Rotation experiment.

Established in 1911, the Cullars Rotation is the oldest continuous soil fertility study in the Southern United States. The best of the Old Rotation treatments were duplicated in the Cullars plots. These plots are located a half mile from the Old Rotation on a Marvyn loamy sand (fine-loamy, siliceous thermic Typic Hapludult). Unfortunately, all pre-1920 Cullars Rotation data were also lost in the 1920 Comer Hall fire. Therefore, the 1916–1919 Old Rotation data could not be resurrected.

The types and levels of actual inputs used on the Old Rotation experiment plots were not recorded, other than to document fertilization treatments. Therefore, all management and input practices on the Old Rotation (i.e. pest control, cultivation, etc.) were assumed to be those recommended by the Alabama Cooperative Extension Service (1983) for use by Alabama growers. Input budgets were constructed by compiling information on 'typical' technologies for each era of the experiment. Information on the quantities of labour used, on the pesticides applied, machinery operations, etc., was drawn from a number of sources, including Cooperative Extension Service budgets, Alabama Experiment Station publications (e.g. White, 1951; Christopher and Roy, 1945), and US Dept. of Agriculture publications (see References for a list of the various USDA sources used).

In most cases, the input series represent recommended technologies for medium sized farms. Indexes of prices paid by producers for input items were used to estimate variable and fixed cost series in years where actual costs were not available (Table 4.3). Cotton lint and seed price data were obtained from statistical abstracts of agricultural prices.

Cotton lint yields comprising the output series are shown as five-year averages in Figure 4.5. In 1896, the no N and winter legume treatment lint yields were 432 kg/ha and 241 kg/ha respectively. The 1955 yield for the 134 kg/ha N treatment was 567 kg/ha. During the past 10 years, output from the plots in continuous cotton have averaged approximately 678 kilograms of lint per hectare on the winter legume treatment, 699 kilograms of lint with 134 kg/ha N, and only 240 kilograms of lint with no N.

External events have affected yield trend. The boll weevil assault on Alabama cotton was documented in 1915. It was claimed that the boll weevil caused 30% reduction in average yields in the state during the period 1910 to 1914 (Hinds and Thomas, 1920). These same scientists credited 315 kilograms per hectare

**Table 4.3** Selected variable and fixed cotton production costs per hectare

| Year | Index prices paid | Total variable US$/ha | Total fixed US$/ha |
|---|---|---|---|
| 1990 | 2419 | 1066.15 | 290.37 |
| 1980 | 1483 | 761.72 | 195.08 |
| 1970 | 526 | 314.09 | 117.33 |
| 1960 | 381 | 600.68 | 10.70 |
| 1950 | 275 | 283.75 | 6.00 |
| 1940 | 153 | 88.77 | 3.36 |
| 1930 | 152 | 118.54 | 2.45 |
| 1920 | 167 | 101.89 | 2.67 |
| 1910 | 102 | 65.60 | 1.63 |
| 1900 | 82 | 52.64 | 1.31 |
| 1896 | 82 | 52.64 | 1.31 |

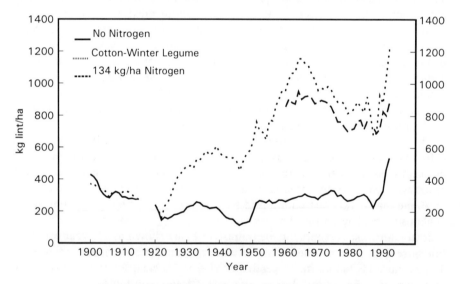

**Figure 4.5** Old rotation yields (five-year averages)

increases in lint to improvements in dusting equipment used to apply pesticides to the boll weevil. Changes in the pesticides applied, machinery used, and improvements in cotton varieties were also seen to affect yield trend.

No herbicides or defoliant cost was included before 1970. Prior to this time, weed control was assumed to be accomplished by hand labour. Fertilizer and lime costs were based on records of rates applied to the plots. Lime costs were estimated based on applied rates of 2.24 Mg/ha every three years.

Except for shifts to tractors from mules and to herbicides from hoes, field operations have remained fairly constant since 1896. Cotton production was powered by animals and humans from 1896–1939. A transition period from animal to tractor power occurred during the period 1940–1955. Since 1955, production has shifted from 2-row (1956–70) to 4-row (1970–85) to 6-row (1986 to present) soil tillage and planting equipment. Marginal improvements were made in ploughing, leveling, bedding and planting equipment. Representative machinery operations for a typical cotton farm consisted of cutting stalks after harvest; flat-breaking the land using a plough; pulverizing the broken land using disk harrows, harrows and/or a drag; bedding; laying off the rows or opening furrows; planting; fertilizing; cultivating; hand thinning; poisoning; harvesting and hauling. Costs of machinery operations and other miscellaneous input costs were estimated using various experiment station bulletins and government publications (Garman, 1932; Davis, 1939; Lanham and Lagrone, 1942; Mahan and Marsh, 1943; Lanham, 1947; Brown, 1950; Robinson, 1951; White, 1951; Mayton and Roy, 1952; Yeager, Belcher and Walkup, 1960; Yeager, 1968; Sutherland et al., 1971, 1974; Ingram and White, 1974). Machinery complements and other input costs from 1970 to present were estimated using Alabama Cooperative Extension Service cotton enterprise budgets.

From 1896–1959 harvesting was accomplished almost exclusively by hand picking, requiring approximately 225 labour hours per hectare. During the 1960s, the cotton picker gradually replaced field hands and family harvesting labour; that dramatically decreased the labour required to produce cotton. For this analysis, it was assumed (and the historic record supports this) that, by 1970, 100% of the cotton was mechanically harvested.

Variable costs included in the analysis accounted for seed, fertilizer, pesticides, defoliation, interest on operating capital, harvesting, ginning, and warehousing, as well as for machinery operation. Fixed costs of operation included tractor and machinery depreciation, interest, insurance and taxes. Estimated returns were to land, management, and owner-operator labour.

Ginning and labour were the other major inputs into the production process. Indexes were used to estimate ginning costs from Extension Service budgets. Budgets were also used to estimate labour rates and wages for the period 1978–1991. Alabama wage rates for 1923–1976 were taken from Agricultural Statistics bulletins. 57% of the US wage rate was used as the Alabama wage rate prior to 1923 (Department of Agriculture, 1899–1919 issues).

## 4.6 RESULTS

Cotton productivity was assessed from three different perspectives. The first perspective is provided by a set of conventional indexes which ignore the negative

## Long-term cotton productivity, 1896–1992

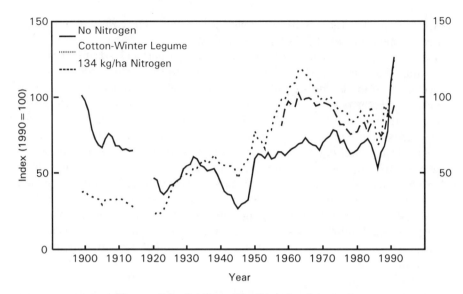

**Figure 4.6** Output index (five-year averages)

external costs associated with erosion and the use of agricultural chemicals. Following this, the effect of assigning a cost to pesticide and erosion externalities is discussed. Finally a TFP index is calculated which allows the performance of the winter legume and 134 kg/ha of Nitrogen treatments to be compared to the performance of the no nitrogen–no legume plot.

When interpreting the first two sets of index figures, it is important to keep in mind that the index series are only relative to an outcome for their respective plot, in the base year of 1990. Time trends can be analysed for each plot, but inferences about the performance differences between treatments can only be made using a third set of indexes. These, relative to no legumes or nitrogen treatment indexes, are presented in the final section of the report. For example, Figure 4.5 shows that yields of the winter legume treatment have exceeded yields on the no N treatment, especially since 1920. Yet the no N output index (Figure 4.6) is substantially greater than the winter legume index in the early years and equals it in the 1990s. This just indicates that 1896–1910 and 1990 were high yield years for the no N plot relative to no N yields in other years. The low index numbers for the early years of the winter legumes plot indicates that yields were low relative only to the later years for this plot. Note that all figures presented in this paper are five-year moving averages.

### 4.6.1 Index without considering external costs

The movements of the yield and output index series (Figures 4.5 and 4.6) are similar. Both Figures show that, on the no N plot, output declined steadily during

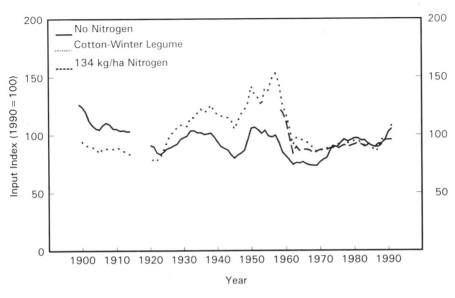

**Figure 4.7** Input index without externalities (five-year averages)

the first 50 years of the experiment. When the Old Rotation experiment was initiated, lint yields on this plot averaged nearly 450 kg/ha. By the 1920–1950 period, yields fell to less than 225 kg/ha. Since that time, output on this plot has recovered slightly.

The output series for the winter legume treatment presents an interesting contrast to the no N plot. The winter legume plot also declined in the early years of the experiment, but then experienced a 40-year period (1920–60) during which yields increased fivefold. This increase has been attributed to higher rates of phosphorus and potassium fertilization, better timing of fertilization, better insect control, and improved varieties. Output has fallen off since the early 1960s, but the 1990 output index is still three times higher than the 1900 index.

The trend on the 134 kg/ha N plot is similar to the movement on the winter legume plot. Output on this plot peaked in the early 1960s and gradually declined into 1990.

The input index series (Figure 4.7) shows less movement than the output index series, suggesting that the total quantity of inputs used has changed relatively less than output, despite the tremendous changes that have been witnessed in the types of chemical and mechanical inputs. The decline in the input index on all three plots from 1950 to 1970 can be attributed to the introduction of the mechanical cotton harvester. Hand picking of cotton is an extremely labour-intensive activity. The appearance of the cotton harvester had the effect of reducing the overall production labour requirement (per hectare) by approximately 70%.

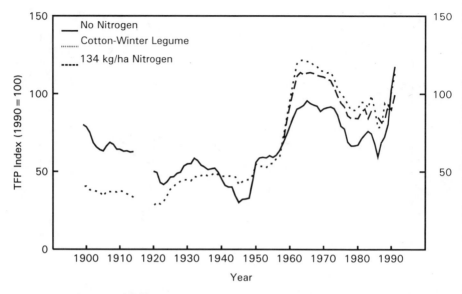

**Figure 4.8** Total factor productivity without externalities (five-year averages)

Three distinct eras of productivity change are exhibited by the TFP series (Figure 4.8). The TFP of the no N plot eroded steadily over the first 50 years of the experiment, bottoming out in the mid-1940s at less than 40% of the 1900 level. The turn-of-the-century decline on the winter legume plot was shorter and more moderate, reaching its low point in 1921 at 70% of the 1900 level. Productivity on all three plots peaked in the middle 1960s, declined during the 1970s and appears to have levelled off in the 1980s. The TFP of all plots in 1991 is greater than when the experiment began. The largest single event affecting productivity was the introduction of the mechanical harvester (1959). The productivity decline of the 1970s is most likely due to the effect of the loss of the DDT insecticide.

### 4.6.2 Total social factor productivity (TSFP)

The Total Social Factor productivity (TSFP) indexes calculated in this section differ from the TFP indexes of the preceding section only in that a cost is assigned to soil erosion and pesticide use. The major categories of external costs of pesticide use are regulatory costs, adverse health effects, and damage to the natural environment. The actual net per-hectare cost (net of external benefits) of these effects is not known. The approach following in this study is to assume external costs equal to 50% of expenditures on herbicides, insecticides, and defoliants. For soil erosion, Ribaudo's (1989) $2.58 per metric ton of soil loss, for annual off-site damage costs in the Southeast US, was multiplied by an estimated average soil loss quantity (calculated as explained below). In reality, soil loss was not the same in all years since it is largely weather dependent. However, since a complete (daily) weather series dating back to 1896 is not available, an average soil loss was assigned for all years.

The Southern Piedmont of the United States has been referred to as one of the 'most severely eroded agricultural areas in the United States' (Trimble, 1974). According to Trimble, the average depth of soil loss over the region was 17.8 centimetres from 1700 to 1970. The most intense erosion occurred from 1860 to 1920, primarily as a result of the cotton culture. Although informal observation leads to the conclusion that erosion has occurred on the Old Rotation, a historical record of erosion does not exist for the site. Past erosion was modelled using the Erosion Productivity Index Calculator simulation programme (Williams, 1990). EPIC simulated 96 years of soil and organic matter loss for the no N and winter legume experimental plots and 35 years of loss for the 134 kg/ha N plot (Tables 4.4, 4.5, 4.6). Ten years of daily wind and rainfall data for the Auburn area were used to generate historic weather conditions. A soil series equivalent to the Old Rotation's Pacolet was used for simulating changes on the three plots. Management constraints were equivalent to those used in 1990.

Average annual soil loss (USLE) over the simulations was estimated to be 22.3 Mg per hectare for the no N treatment; 13.6 Mg per hectare on the winter

**Table 4.4** Universal soil loss and other environmental factors estimated using EPIC; five-year average values for continuous cotton – no nitrogen

| Year | Yield kg/ha | USLE Mg/ha | Rain cm | OM loss kg/ha | Runoff cm |
|---|---|---|---|---|---|
| 5 | 608.97 | 23.63 | 166.16 | 24.33 | 20.78 |
| 10 | 587.98 | 15.52 | 140.94 | 16.86 | 12.32 |
| 15 | 599.18 | 13.79 | 148.60 | 14.48 | 13.08 |
| 20 | 532.60 | 15.45 | 160.86 | 14.75 | 14.62 |
| 25 | 436.89 | 23.37 | 178.56 | 17.92 | 19.60 |
| 30 | 406.74 | 16.97 | 153.46 | 15.90 | 17.63 |
| 35 | 369.56 | 25.46 | 173.10 | 17.89 | 19.54 |
| 40 | 387.83 | 21.78 | 150.54 | 13.43 | 17.12 |
| 45 | 326.26 | 23.03 | 143.68 | 12.70 | 13.09 |
| 50 | 381.79 | 21.25 | 151.70 | 11.62 | 15.26 |
| 55 | 358.13 | 16.43 | 135.44 | 9.63 | 11.39 |
| 60 | 322.09 | 39.56 | 180.13 | 15.53 | 23.68 |
| 65 | 354.61 | 24.73 | 160.10 | 11.62 | 15.38 |
| 70 | 327.13 | 23.38 | 159.93 | 11.60 | 15.67 |
| 75 | 394.91 | 25.22 | 157.51 | 9.93 | 20.21 |
| 80 | 341.91 | 28.15 | 168.16 | 11.13 | 20.37 |
| 85 | 410.14 | 25.22 | 168.01 | 10.31 | 19.39 |
| 90 | 384.65 | 14.17 | 140.53 | 7.02 | 11.09 |
| 95 | 339.70 | 25.30 | 153.13 | 9.21 | 18.91 |

**Table 4.5** Universal soil loss and other environmental factors estimated using EPIC; five-year average values for cotton – Winter Legumes

| Year | Yield kg/ha | USLE Mg/ha | Rain cm | OM loss kg/ha | Runoff cm |
|---|---|---|---|---|---|
| 5  | 859.17  | 15.17 | 166.16 | 17.26 | 20.65 |
| 10 | 1079.75 | 10.70 | 140.94 | 12.82 | 11.89 |
| 15 | 1028.76 | 9.31  | 148.60 | 11.30 | 12.99 |
| 20 | 1243.54 | 10.49 | 160.86 | 11.70 | 14.41 |
| 25 | 1121.01 | 13.64 | 178.56 | 12.92 | 19.57 |
| 30 | 878.80  | 9.73  | 153.46 | 11.89 | 17.53 |
| 35 | 1004.57 | 17.18 | 173.10 | 14.43 | 19.46 |
| 40 | 1105.40 | 10.37 | 151.79 | 10.47 | 17.00 |
| 45 | 1107.57 | 14.51 | 143.68 | 13.70 | 13.08 |
| 50 | 1173.36 | 10.07 | 151.70 | 12.66 | 15.40 |
| 55 | 941.49  | 10.61 | 135.44 | 11.53 | 11.62 |
| 60 | 1071.95 | 24.57 | 175.77 | 15.25 | 23.67 |
| 65 | 1243.20 | 18.32 | 151.69 | 11.84 | 15.12 |
| 70 | 684.81  | 16.06 | 157.72 | 10.73 | 16.08 |
| 75 | 969.83  | 18.10 | 157.51 | 9.69  | 20.21 |
| 80 | 1123.76 | 13.68 | 168.16 | 10.36 | 20.23 |
| 85 | 1281.19 | 13.80 | 168.01 | 11.27 | 19.10 |
| 90 | 881.51  | 7.84  | 140.53 | 8.23  | 11.15 |
| 95 | 1103.16 | 15.24 | 153.13 | 10.50 | 18.98 |

**Table 4.6** Universal soil loss and other environmental factors estimated using EPIC; five-year average values for cotton – 134 kg/ha Nitrogen

| Year | Yield kg/ha | USLE Mg/ha | Rain cm | OM loss kg/ha | Runoff cm |
|---|---|---|---|---|---|
| 5  | 1531.80 | 22.23 | 166.16 | 22.30 | 21.25 |
| 10 | 1197.62 | 14.55 | 140.94 | 16.68 | 12.55 |
| 15 | 1498.07 | 13.13 | 148.60 | 15.12 | 13.35 |
| 20 | 1519.77 | 14.16 | 160.86 | 15.49 | 14.92 |
| 25 | 1423.03 | 19.41 | 178.56 | 17.90 | 20.02 |
| 30 | 1138.79 | 16.95 | 153.46 | 16.81 | 17.92 |
| 35 | 1191.52 | 20.29 | 173.10 | 18.86 | 19.90 |

legume treatment; and 17.47 Mg per hectare for the 134 kg/ha N treatment. Organic nitrogen lost in sediment was 13.4 kg/ha for cotton with no nitrogen; 12 kg/ha for with a winter legume and 17.6 kg/ha for cotton with 134 kg/ha of nitrogen.

The inclusion of external costs did not significantly affect productivity trends in any of the treatments. The no N plot indexes are not changed at all; the input indexes (Figure 4.9) on the other two plots increased by an average of about 6%. Total Factor Productivity on the legume and 134 kg/ha of nitrogen plots

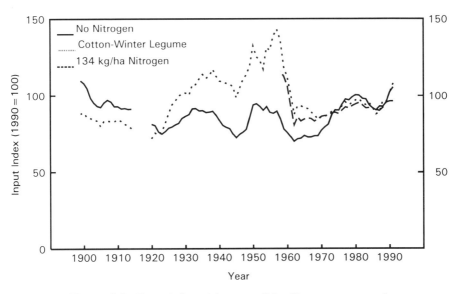

**Figure 4.9** Input index with externalities (five-year averages)

decreased by 4 and 6% respectively (Figure 4.10). The main conclusions of the previous section are therefore unaffected. Thus, total input use remains relatively stable over time whether or not a cost is assigned to externalities and significant productivity growth occurs whether measured as conventional TFP or as TSFP.

### 4.6.3 Productivity effects of winter legume and chemical nitrogen

It is not possible to assess the relative performance of the three plots using either of the index series discussed above. However, the input use, output use and productivity of the winter legume and 134 kg/ha N treatments relative to the no N treatment can be calculated as indexes (Cooke and Sundquist, 1991). Where:

$$I(X)_{ps} = 100 \, e^{[\Sigma_i .5(S_{ipt}/S_{i1t}) \ln(X_{ipt}/X_{i1t})]} \tag{6.4}$$

$$I(Y)_{pt} = 100 \, e^{[\Sigma_j .5(W_{jpt}/W_{j1t}) \ln(Y_{jpt}/Y_{j1t})]} \tag{6.5}$$

$$TFP_{pt} = I(Y)_{pt}/I(X)_{pt} \tag{6.6}$$

and where

$S_{ipt}$ = the cost share of input $i$ for plot $p$ in $t$,
$S_{i1t}$ = the cost share of input $i$ for plot 1 in $t$,

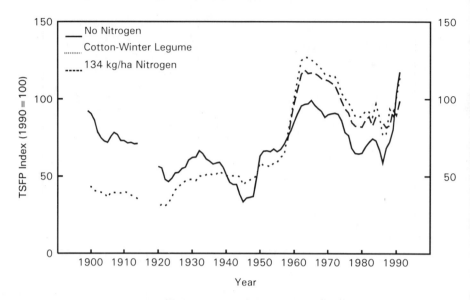

**Figure 4.10** Total social factor productivity (five-year averages)

$X_{ipt}$ = the quantity of input $i$ used on plot $i$ in $t$,
$X_{i1t}$ = the quantity of input $i$ used on plot 1 in $t$,
$S_{ipt} = r_{it}X_{ipt}/\Sigma r_{it}X_{ipt}$ where $r_{it}$ is the price of input $i$, and
$W_{jpt} = p_{jt}Y_{jpt}/\Sigma p_{jt}Y_{jpt}$ where $p_{jt}$ is the price of output $j$.

The index numbers derived from equations (4.4)–(4.6) are not linked and are not relative to a base year as has been the case in the preceding discussion, but they are calculated relative to the no nitrogen–no legumes *base plot*; that is the index value for the no nitrogen-no legumes plot is equal to 100 in all years.

The relative output indexes merely confirm what was shown in Figure 4.5; outputs of the no N and winter legume plots were similar during the early years of the experiment, but then diverged. Yield and output on the winter legume and 134 kg/ha N plots were similar throughout the period 1955–1991.

The input series (Figure 4.11), as expected, shows that the winter legume and 134 kg/ha N plots used more input than the no N plot. However, much of this apparent input intensity is related to increased harvest costs, which were due to higher yields on the legumes and nitrogen plots. More harvest labour is needed, and ginning costs are higher, on a hectare that yields four bales than on a hectare that yields two bales of cotton.

The most informative inter-plot comparisons are obtained from the productivity measures, TFP (Figure 4.12) and TSFP (not shown). The winter legume plot relative TFP hovers near 100 until 1920, climbs to 200 in the 1940s, declines in the 1950s through the 1970s, and climbs back to 200 in the 80's and 90's. The

Results 59

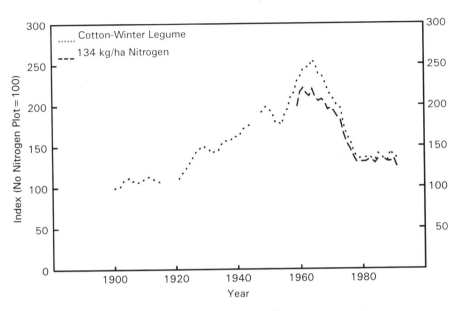

**Figure 4.11** Relative input index (five-year averages)

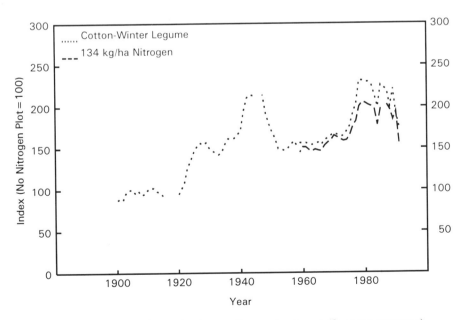

**Figure 4.12** TFP alternative to no nitrogen treatment (five-year avearges)

134 kg/ha N plot relative TFP series follows a similar pattern, but is slightly below the winter legume plot in all years.

One of the most interesting findings is that there is a greater difference in relative TSFP than relative TFP. The difference is 6–8% in most years on both plots. This implies that accounting for externalities enhances the productivity advantage of the winter legume and chemical nitrogen plots. In other words, this appears to be a case where the low input system (no N), has less desirable environmental consequences. This comes about through the decrease in soil erosion brought about through the higher biomass production of the winter legume and 134 kg/ha N plots.

## 4.7 CONCLUSIONS

This paper used index numbers to examine trends in productivity of three of the thirteen cotton experimental treatments contained in Alabama's Old Rotation. Test plot data were used to construct three different sets of index numbers: TFP which accounts for all direct production costs, but which does not consider production externalities; TSFP which accounts for all direct production costs as well as external costs of soil erosion and pesticide use; and productivity relative to a base plot.

The most difficult part of the exercise was constructing a complete input series for the 1896–1991 period. The research team's subjective judgement entered into the determination of the representative technology in a number of ways. In years in which both USDA and Alabama Cooperative Extension Service budgets were available, judgements were made as to which technology was most representative of the actual practices used. Where no information was available, prices paid and received indexes were used to reconstruct budgets.

Viewed from the 97-year perspective of the Old Rotation experiment, all three plots fulfil at least one criterion required for a system to be sustainable. That is, output per unit of input is higher in 1991 than in 1896, even when externalities are valued. The final observations that can be made about the systems analysed are:

1. None of the systems shows a linear trend in output or TFP over the life of the experiment. Productivity cycles are present in all three systems, despite the positive overall trend. An important focus of future research will be to attempt to explain whether these cycles are related to weather, technology, or changes in the resource base. Casual observation suggests that weather related effects are dominant.
2. The system which has neither an organic or a chemical source of added nitrogen (plot 6) is less productive than the other two systems. This system compares even more poorly when externality costs are assigned.
3. Organic and chemical sources of nitrogen have similar productivity impacts.

4. Soil erosion and pesticide externalities have a modest effect on measured productivity.
5. The most dramatic single event to affect productivity was the introduction of the mechanical cotton picker. The impact of this technology is powerful enough to offset the effect of many other changes in the system.

# 5

# Extrapolating Trends from Long-term Experiments to Farmers' Fields: The Case of Irrigated Rice Systems in Asia

K. G. Cassman and P. L. Pingali

*International Rice Research Institute, Philippines*

## 5.1 INTRODUCTION

Long-term field experiments provide a vehicle for scientific investigation of structure-function relationships that govern the long-term viability of a specified cropping system. The key issues are: does imposition of a specified cropping pattern and management regime alter the biological, physical, and chemical properties of the soil resource base; and if changes occur, how do such alterations affect crop growth, input use efficiency, and yield? Maintaining long-term field studies, however, is a costly proposition and thus the value of such endeavours must be justified in relation to other research opportunities which can also lead to increased output, productivity, and preservation of the natural resource base. Most important is whether information generated from a long-term field experiment conducted at a research station can be used to foresee constraints to future productivity in farmers' fields where similar cropping systems are practised: constraints that would be difficult or impossible to detect in short-duration field studies or by other research approaches.

---

*Agricultural Sustainability: Economic, Environmental and Statistical Considerations.*
Edited by V. Barnett, R. Payne and R. Steiner © 1995 John Wiley & Sons Ltd

The purpose of this paper is to make a comparative analysis of productivity trends in rice systems of tropical and subtropical Asia based on

1. aggregate data at the national level,
2. long-term farm monitoring studies and district-level data, and
3. long-term experiments on continuous, irrigated rice systems that were established well before such systems were widely adopted by rice farmers.

Our goal is to identify constraints to further productivity growth at each level of aggregation, to discuss the potential causes of these constraints, and to evaluate the relevance of data from long-term experiments as a tool for efficient targeting of research on mitigation or avoidance of such constraints. To achieve this goal requires common performance measures, or 'common parameters', that are relevant across different levels of aggregation.

## 5.2 LINKING STRUCTURE TO FUNCTION OF CROPPING SYSTEMS

System function, or performance, can be quantified in many ways. Economists are concerned with the conditions for profit maximization and the discounted value of future returns from current investment and technology choices. Ecologists look at stability and resilience in terms of material and energy balances, interactions among trophic levels, regulating feedback mechanisms, and biodiversity. Agronomists seek to harness solar, soil, and water resources to optimize crop yields in relation to input requirements. Farmers are concerned with maximizing their return on labour, land, or capital depending on the socio-economic and policy environments in which they operate.

Two measures of system performance that link the perspectives of farmers and scientists are the efficiency with which inputs are utilized to produce the desired output and the total output which is determined by yield per unit area. A broad economic index of output/input efficiency is *total factor productivity* (TFP), which is the ratio of total output value (i.e. grain and harvested biomass or forage) to the total cost of all inputs used to produce the crop, including labour, animal work, fossil fuels, organic and inorganic nutrient applications, pesticides, irrigation water, and seed. Lynam and Herdt (1989) suggest 'the appropriate measure of output by which to determine sustainability at the crop, cropping system, or farming system level is total factor productivity; a sustainable system has a non-negative trend in TFP over the period of concern'.

Changes in TFP over time may result from changes in the socio-economic and biophysical environment. Evaluating TFP trends after specifying constant prices for inputs and outputs eliminates fluctuations due to changes in market conditions but does not account for changes in the policy environment or changes in the biophysical resource base which may influence the quantity and efficiency of inputs used by farmers. Thus, monitoring of the TFP trend-line based on constant prices indicates that the efficiency of input use has changed but does not explain why the change is occurring.

In cases where a decline in TFP results from a reduction in the quality of the biophysical resource base to support crop growth, it is the job of biological and physical scientists to quantify the effects of the crop production system on soil quality and pest pressure, and to understand the processes responsible for these changes. Partial factor productivities based on the key biophysical constraints to crop production in a given environment can be monitored to indicate the trend-line for biological efficiency. Relevant indices include output/input ratios for nutrients, grain yield per unit water input as rainfall and irrigation, or an energy ratio based on the calorific value of all outputs and inputs. When a partial factor productivity index changes over time in a long-term experiment with input levels and crop management practices held constant, such a change is caused by an alteration of at least one component of the natural resource base and these changes can be quantified through measurement of the appropriate biophysical parameters.

We shall use TFP, partial factor productivities, and yield as the basis for evaluating performance of rice systems at different levels of aggregation.

## 5.3 REGIONAL AND NATIONAL TRENDS IN PRODUCTIVITY AND YIELD

Although rice output and yield continue to increase throughout most of Asia, recent trends indicate yield stagnation and declining factor productivity which are cause for concern. The annual rate of growth in aggregate rice output increased from 2.1% from 1955–64 to 3.3% from 1964–1981 due to the introduction and spread of high yielding modern rice varieties in irrigated and favourable rain-fed lowland areas (Dalrymple, 1986; Herdt and Capule, 1983). Expansion of rice area contributed about one-third of the output growth in the 1960s and one-fifth in the 1970s. In the 1980s, however, area expansion virtually halted. Most of the increase in rice output from 1964–81 resulted from increased yield, but growth rates in yield have declined sharply in the 1980s (Table 5.1). For Asia as a whole, yield growth rates decreased from 2.6% per annum in the 1970s to 1.5% from 1981–88. Yield increases in the 1980s were slowest in South Asia (excluding India), and substantial declines in yield growth also occurred in China and South-east Asia.

Concurrent with the decrease in yield growth rate, aggregate data from some countries suggest declining partial factor productivity for certain inputs (Rosegrant and Pingali, 1994). In Indonesia from 1976–86, for example, total rice production increased by 70%, mostly due to increases in yield, whereas estimates of nitrogen(N) fertilizer use on rice increased by 440% (IRRI, 1991) (Figure 5.1).

**Table 5.1** Annual growth rates in rice area, production, and yield, Asia, 1961–88

| Region | 1961–72[a] (%) | 1972–81 (%) | 1981–88 (%) | 1961–88 (%) |
|---|---|---|---|---|
| Asia[a] | | | | |
| Area | 1.00 | 0.66 | 0.02 | 0.59 |
| Production | 3.30 | 3.32 | 1.52 | 3.08 |
| Yield | 2.28 | 2.64 | 1.50 | 2.47 |
| South Asia[b] | | | | |
| Area | 1.37 | 1.24 | −0.07 | 0.86 |
| Production | 2.00 | 3.14 | 1.01 | 2.21 |
| Yield | 0.62 | 1.88 | 1.08 | 1.33 |
| South-east Asia[c] | | | | |
| Area | 0.68 | 1.45 | 0.62 | 0.87 |
| Production | 2.76 | 3.99 | 2.24 | 3.31 |
| Yield | 2.07 | 2.50 | 1.61 | 2.42 |
| China[a] | | | | |
| Area | 2.15 | −0.66 | −0.62 | 0.29 |
| Production | 4.01 | 2.54 | 1.81 | 3.23 |
| Yield | 1.83 | 3.23 | 2.45 | 2.93 |
| India | | | | |
| Area | 0.62 | 0.97 | 0.05 | 0.60 |
| Production | 1.92 | 1.03 | 3.27 | 2.48 |
| Yield | 1.29 | 2.05 | 3.21 | 1.87 |

[a]For Asia and China, first period trends are for 1964–72. Asia includes South Asia, South-east Asia, China, and India.
[b]South asia includes Bangladesh, Nepal, Pakistan, and Sri Lanka, excluding India.
[c]South-east Asia includes Burma, Indonesia, Kampuchea, Laos, Malaysia, Philippines, Thailand, and Vietnam.

Source of data: For China, State Statistical Bureau, Ministry of Agriculture. For all other countries, FAO data tapes.

Diminishing returns from higher rates of N application and favourable subsidies that may have encouraged inefficient use of fertilizer were likely contributors to the decline in N factor productivity. However, the magnitude of the decrease in N fertilizer factor productivity, from 75 to 28 kg of grain output per kg applied N, raises the question of whether degradation of the paddy resource base due to the imposition of continuous, irrigated rice monocropping has also contributed to declining N output/input efficiency.

## 5.4 FARM-LEVEL TRENDS IN PRODUCTIVITY AND YIELD

It is difficult to assess trends in farm performance and the technical efficiency of farmers over time using national average yield time series because these data are

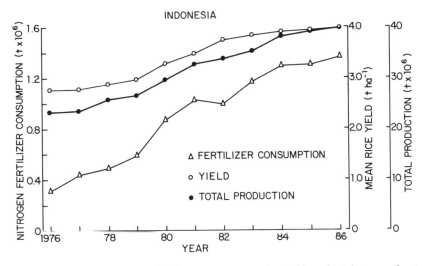

**Figure 5.1** Trends in nitrogen fertilizer use, average rice yield, and total rice production in Indonesia from 1976 to 1986 (IRRI, 1991)

based on the pooling of heterogeneous rice-growing environments (Herdt 1988; Barker et al., 1985). Monitoring of yield trends and farming practices in provinces or districts where the agroclimatic environment is more uniform allows a more accurate assessment of productivity trends and technical efficiency, especially when coupled with data from a nearby experiment station for comparison. In the following section, trends in farm-level productivity are assessed in selected irrigated rice areas. Two intensively cropped rice domains were chosen in each of two countries, the Philippines and India. Yield trends and trends in partial productivity indexes such as yield per kg applied N and yield per unit labour are compared with trends in TFP.

Long-term, farm-level data that can be used to discern trends in productivity are rare throughout Asia. The exception are the long-term farm monitoring data from Laguna and Central Luzon provinces in the Philippines which were collected by IRRI's Social Sciences Division from 1966–1990. In the case of India, rice-growing domains were selected where irrigated rice systems have predominated since the late 1960s and district-level data can be used to detect productivity trends. The Indian locations are the Ludhiana District, Punjab and the Krishna District in Andhra Pradesh.

Yield trends within these selected domains indicate patterns that are similar to the aggregate national and regional trends (Table 5.1) in rice output: there was a dramatic increase in yields following the rapid adoption of modern rice varieties and management practices in the 1960s, yields continued to increase in the second decade after adoption, and stagnant or declining yield trends are indicated in the most recent decade. Based on the farm monitoring data from

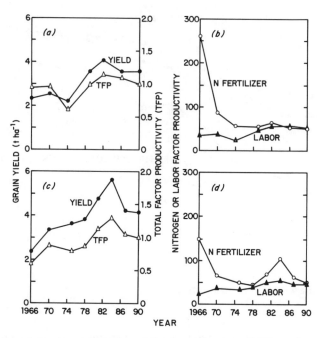

**Figure 5.2** Trends in rice yield and total factor productivity (TFP), and partial factor productivity for N fertilizer (kg paddy/kg applied N) and labour (kg paddy/person-day) achieved by farmers in Central Luzon (a and b) and Laguna (c and d) provinces, Philippines

the Philippines, average wet season (WS) rice yields were approximately 2.5 t/ha in 1966, and had increased to 4.2 t/ha in Central Luzon and 4.7 t/ha in Laguna by the early 1980s (Figure 5.2a, c). Since then, however, there has been a gradual decline in yields so that grain yield was 0.5 t/ha lower in 1990 than in the early 1980s in both domains.

In Ludhiana, Punjab, where an intensive rice–wheat double crop system is practiced, average rice yields rose from 1.8 t/ha in 1970 to 4.0 t/ha by 1980 and have remained relatively constant since (Figure 5.3a). Investment in irrigation and adoption of modern rice technology have combined to produce similar trends in the delta areas of Southern India where intensive, continuous rice monoculture is practiced. Yield trends in Krishna District, a delta district of Andhra Pradesh, indicate yields remain on a positive trend but are growing slowly (Figure 5.3c).

Yield trends alone do not tell the complete story because farmers continually modify their management practices as new technologies become available and in response to government policies and markets. The Central Luzon farmers, for example, have changed from a single rice crop per year in the 1960s to a continuous rice double crop system since the mid-1970s. They switched from

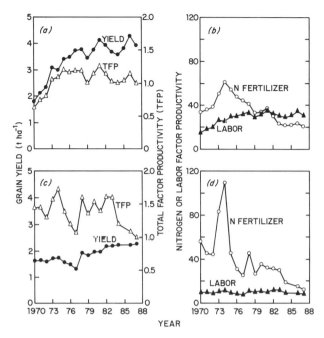

**Figure 5.3** Trends in rice yield and total factor productivity (TFP), and partial factor productivity for N fertilizer (kg paddy/kg applied fertilizer) and labour (kg paddy/person-day) achieved by farmers in Ludhiana, Punjab (a and b) and Krishna District, Andhra Pradesh (c and d), India

transplanting, which requires small quantities of seed planted in seedbeds, to a direct wet-seeded establishment technique that requires threefold greater seeding rates (Table 5.2). The total labour requirement increased initially due to more labour for harvesting as yields and application of inputs increased, followed by decreasing labour inputs as rotary cultivators became available for land preparation and direct seeding reduced labour needs compared with transplanting. Pesticide and N fertilizer inputs increased rapidly up to the early 1980s, but input rates have remained relatively constant in the past decade.

Similar trends in input use are apparent for the Laguna farmers except that transplanting remains the primary method of crop establishment. Intensification of cropping systems also occurred in the two Indian districts with a rapid increase in mechanization in the Punjab and large investments in irrigation in both the Punjab and Andhra Pradesh during the past two decades. As discussed previously, assessment of changes in productivity as cropping systems evolve requires the quantification of trends in total and partial factor productivities.

Based on 1990 prices for inputs and outputs in each year in the Philippines, trends in TFP indicate that productivity increased steadily until the early 1980s, followed by a slight decline through 1990 in both Central Luzon and Laguna

**Table 5.2** Mean input levels used by farmers on wet season rice in Central Luzon, 1966–90

| Year | No. of farmers | N rate (kg/ha) | Pesticide use* (kg/ha) | Seed rate (kg/ha) | Total labour (d/ha) | Tractor use (d/ha) |
|---|---|---|---|---|---|---|
| 1966 | 91  | 9  | 0.1 | 45  | 44 | 0.1 |
| 1970 | 62  | 29 | 0.2 | 52  | 47 | 0.2 |
| 1974 | 58  | 39 | 0.5 | 49  | 62 | 0.3 |
| 1979 | 146 | 63 | 0.8 | 115 | 50 | 0.7 |
| 1982 | 136 | 63 | 1.0 | 129 | 48 | 0.8 |
| 1986 | 120 | 67 | 0.7 | 184 | 42 | 0.8 |
| 1990 | 107 | 70 | 0.9 | 169 | 42 | 1.3 |

*Based on kg active ingredient/ha.

(Figures 5.2a, c). TFP trends in Ludhiana follow a similar pattern to those in the Philippines (Figure 5.3a), while there has been a sharp decline in TFP since 1984 in Andhra Pradesh (Figure 5.3c).

Partial factor productivity for labour followed a pattern similar to the trend-line in TFP in all cases except Andhra Pradesh where relatively little mechaniza-

**Table 5.3** Production function estimates: Laguna and Central Luzon, Philippines, 1966–90

|  | Laguna | Central Luzon |
|---|---|---|
| Intercept: (1966–1974) | 2111 | 1663 |
| Nitrogen/ha | 14.5 | 19.1 |
|  | (4.95) | (5.24) |
| (Nitrogen/ha)$^2$ | −0.04 | −0.06 |
|  | (2.58) | (2.39) |
| Labour use/ha | 3.2 | 2.8 |
|  | (1.48) | (1.12) |
| Tractor days/ha | 4.4 | 7.1 |
|  | (0.14) | (0.19) |
| Seed/ha | 3.6 | 2.2 |
|  | (3.61) | (3.03) |
| Insecticides/ha | — | 292 |
|  |  | (5.95) |
| Dummy: 1979–82 | 958 | 606 |
|  | (7.64) | (4.50) |
| Dummy: 1986–90 | 696 | 186 |
|  | (4.23) | (1.14) |
| F-value | 35.86 | 44.96 |
| Degrees of freedom | 692 | 711 |

Figures in parenthesis are *t*-values.

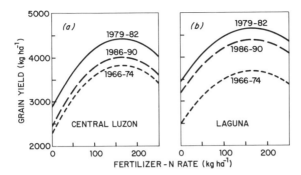

**Figure 5.4** Predicted rice yield response to N fertilizer inputs in three time periods for farmers in two Philippine provinces, based on production functions that hold input levels constant for inputs other than N (see Table 5.3)

tion has occurred (Figures 5.2 and 5.3). In the Philippines, grain yield per unit applied N decreased markedly from 1966 to the early 1970s, then remained relatively constant at about 50 kg grain per kg N input (Figures 5.2b, d). A steady decrease in productivity from applied N is also apparent in the Indian districts since the mid-1970s (Figures 5.3b, d).

Comparison of the direct effect of N rate on grain yield in different time periods, however, requires that inputs other than N be held constant because input substitution and adoption of new technology can affect the response to N. We avoided these confounding effects by estimating a production function in which the rates of all inputs other than N were held constant while examining the yield response to varying rates of fertilizer N inputs in different time periods. Production functions were estimated for Central Luzon and Laguna using cross-section time series data from 1966–1974, 1979–1982 and 1986–1990 which included an intercept-shift dummy variable for the last two time periods. Regression coefficients for these production functions are provided in Table 5.3.

With constant input levels for labour, tractor use, seed rate, etc. specified at the sample mean for the entire data set, the quadratic response to applied fertilizer N shifted over time (Figures 5.4a,b). Maximum N use efficiency is indicated in the 1979–1982 period in both domains with a substantial decrease in efficiency in the latest period which occurred despite little change in the mean rate of fertilizer N inputs from 1979–1990 (Table 5.2).

## 5.5 YIELD TRENDS IN LONG-TERM EXPERIMENTS ON INTENSIVE RICE SYSTEMS

When the first prototypes of modern rice varieties were developed in the early 1960s, IRRI agronomists foresaw the revolutionary impact this new rice plant would have on cropping intensity and input use. Thanks to their vision,

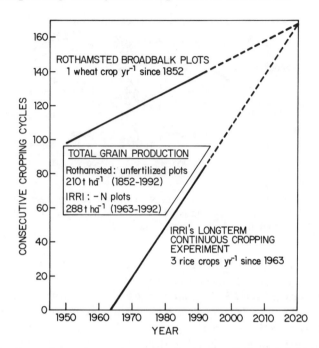

**Figure 5.5** Comparison of cropping intensity in the Rothamsted Broadbalk long-term experiment, and in the Long-Term Continuous Cropping experiment at IRRI

a continuous rice cropping system was established on a 1 ha field at the IRRI farm in 1963. These plots produced 13 rice crops in the five-year period from 1963–1967 (IRRI, 1967), and have supported three rice crops per year since 1968—except in 1991 when only 2 crops were possible when the irrigation supply was interrupted in May and June for refurbishing the pump and well-shaft that supply water to the field. The *Long-Term Continuous Cropping Experiment* (LTCCE) was initiated in this 1 ha field in 1968. Fertilizer N main plots and varietal subplots have been imposed on the same treatment plots in each crop cycle since. The 1993 dry season (DS) rice crop is the 89th consecutive rice crop at this site, and the LTCCE represents the most intensive food crop system under long-term study in the world (Figure 5.5). Today, nearly 30 years after the establishment of these continuously cropped plots at IRRI, triple crop rice systems are common in several lowland areas of tropical Asia where year-round irrigation is available. Where triple cropping is not practiced, double crop continuous rice systems predominate on irrigated land.

Other long-term experiments on continuous double crop irrigated rice systems were established at IRRI from 1964–1966, in 1968 at several research stations elsewhere in the Philippines, and in 1972–73 at two sites in India as part of a project on *Long-Term Fertilizer Experiments* initiated by scientists from the Indian Agricultural Research Institute (Nambiar and Gosh, 1984).

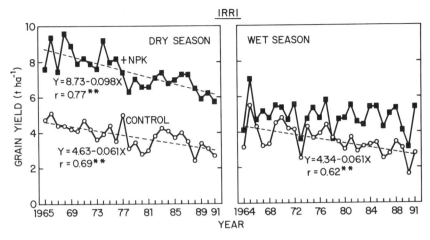

**Figure 5.6** Yield trends of the highest yielding variety in the Long-Term Fertility Experiment conducted at the IRRI Research Farm in Laguna Province, Philippines since 1964 in treatments that receive complete nitrogen (N), phosphorus (P), and potassium (K) inputs in each crop cycle (+NPK), and in the control treatment without fertilizer- nutrient inputs

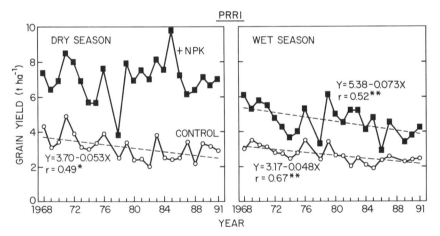

**Figure 5.7** Yield trends of the highest yielding variety in the Long-Term Fertility Experiment conducted at the PhilRice Research Station in Central Luzon, Philippines since 1968 in treatments that receive complete nitrogen (N), Phosphorus (P), and potassium (K) inputs in each crop cycle (+NPK), and in the control treatment without fertilizer-nutrient inputs

It was not until the early 1980s, however, that declining yield trends were recognized at several sites. Flinn *et al.* (1982) published the first report of these trends based on the highest yielding entry of 28 elite lines grown at five different N rates, both in the WS and DS on the same field at IRRI. The germplasm used in

**Table 5.4** Summary of yield trends from long-term experiments with intensive irrigated rice double or triple crop monoculture systems conducted in the Philippines and India. In all cases, yield trends are based on performance of the best cultivars included in each experiment

| Experiment/Site | Period | Season | Zero N or −NPK control (kg/ha/yr) | +N or full NPK | Reference or Source |
|---|---|---|---|---|---|
| LTFE/IRRI | 1964–91 | DS | −60 | −100 | IRRI APPA Div. |
|  |  | WS | −60 | NS[†] |  |
| LTFE/PRRI | 1968–91 | DS | −50 | NS | IRRI APPA Div. |
|  |  | WS | −50 | −70 |  |
| LTFE/BRIARC | 1968–91 | DS | NS | NS | IRRI APPA Div. |
|  |  | WS | NS | NS |  |
| LTCCE/IRRI | 1968–91 | DS | −130 | −130 | IRRI APPA Div. |
|  |  | EWS | −100 | −70 |  |
|  |  | LWS | −70 | −70 |  |
| N Response/IRRI | 1966–80 | DS | −130 | −160 | Flinn et al., 1982 |
|  |  | WS | −130 | −145 |  |
| N Response/PRRI | 1968–80 | DS | NS | NS | Flinn and De Datta, 1984 |
|  |  | WS | NS | NS |  |
| N Response/BRIARC | 1968–80 | DS | NS | NS | ,, ,, |
|  |  | WS | −130[‡] | −130[‡] |  |
| N Response/VRES | 1968–80 | DS | −120[‡] | −120[‡] | ,, ,, |
|  |  | WS | NS | NS |  |
| ICAR-LTFE/ Hyderabad | 1972–81 | DS | −110[§] | −240[§] | Nambiar and Ghosh, 1984 |
|  |  | WS | NS | −120[§] |  |
| ICAR-LTFE/ Bhubaneswar | 1972–81 | DS | NS | NS[§] | ,, ,, |
|  |  | WS | NS | −110[§] |  |

[†]NS = no significant yield trend, either positive or negative.
[‡]Slope of the yield decline trend line did not differ significantly among N rates. Authors reported the mean rate of decline.
[§]Yield trend estimated from a Figure 23 and 24 in cited publication.

this experiment represented the most promising lines available at each point in time, and many would later become IRRI varieties. As in all long-term experiments, certain management practices were consistently imposed as treatments, in this case N-rates, while non-treatment operations such as pest management followed recommended practices which changed over time. Other nutrients such as P, K and Zn were applied as needed or at recommended levels. Flinn et al. (1982) state, however, that 'changes in crop management, if anything, would have slowed down the long-term yield declines'. In subsequent analyses of similar *N Response Experiments* conducted by IRRI researchers at what is now the

Philippine Rice Research Institute (PRRI), the Bicol Region Integrated Agricultural Research Center (BRIARC) in Camarines Sur, and the Visayas Rice Experiment Station (VRES) in Iloilo, significant negative yield trends were also found at two of these three sites (Flinn and De Datta, 1984). In 1988, the long-term *N Response Experiments* were terminated.

Other long-term experiments in the Philippines continue. These include the LTCCE at IRRI, discussed earlier, with three crops per year since 1963 and four N-rate main plot treatments applied to the same plots in each crop cycle since 1968, with six varieties as subplots and four replicate blocks, and the *Long-Term Fertility Experiments* (LTFE) with factorial treatments combining various NPK fertilizer input levels and three varieties, established in 1964 at IRRI, and in 1968 at PRRI and BRIARC. In each of these experiments, IR8 was included as one of the varieties from 1968–1991, while the other varieties used in any given year represented the highest yielding, most disease and insect resistant lines available. As in the *N Response Experiments*, insects and weeds were controlled as required, and limiting nutrients other than treatment variables were applied as needed.

At the LFTE sites, declining yield trends for the highest yielding variety in the complete NPK input treatment are significant at two of the three sites: at IRRI in the DS (Figure 5.6) and in the WS at PRRI (Figure 5.7). In the triple crop system of the LTCCE at IRRI, declining yield trends with the best cultivars are apparent in all three cropping seasons, both in treatments with the highest rate of N inputs as well as in the control treatment without N inputs. The pattern of declining yields both in treatments with NPK input levels that were considered to be optimal at the outset of the experiment and in control plots without nutrient inputs is consistent in most long-term experiments where declines in yield are observed (Table 5.4). Also notable is the absence of any long-term experiment with continuous, irrigated rice in which there is a positive yield trend despite the continual replacement of older varieties with improved varieties.

## 5.6 CAUSES OF THE YIELD DECLINE AND N FACTOR PRODUCTIVITY

Weather data summarized by Flinn *et al.* (1982) and Flinn and De Datta (1984) found that the yield decline was not linked to changes in solar radiation. Likewise, their statistical analyses isolated the influence of viral diseases, typhoons and bacterial leaf blight and leaf streak, and yet still found a significant linear yield decline. These authors proposed several factors as potential contributors to the yield declines: zinc dificiency and boron toxicity at IRRI, and at all sites they suggested the possibilities of other nutrient deficiencies or imbalances; increasing severity of insect pests and their transmitted viral diseases; and greater susceptibility to lodging caused by stem rot and sheath blight infection. To date, however, there has been no documentation of a direct cause and effect relationship between any of these factors and the yield decline.

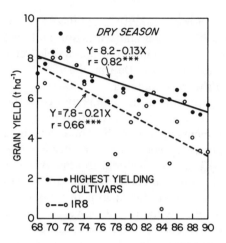

**Figure 5.8** Dry season yield trends of the highest yielding varieties and IR8 with inputs of 150 kg N ha$^{-1}$ in each crop cycle of the Long-Term Continuous Cropping Experiment conducted at the IRRI Research Farm

In the LTCCE, the contribution of increasing disease pressure is evident from the marked fluctuations in yield about the trend-line for IR8 which became highly susceptible to grassy stunt and tungro virus diseases and their vectors (Figure 5.8). This pattern reflects the sporadic nature of diseases like tungro for which severity and yield loss depend on the vector population, availability of the virus which is non-persistent, the stage of the rice plant when infection occurs, and the timing of all these factors (Ou, 1985). Although yield of the highest yielding cultivars in each year also declines steadily, both the rate of decline and variation about the trend-line is considerably less than for IR8. The greater yield and yield stability of the highest yielding varieties reflects the incorporation of resistance to the vectors of viral diseases in the more recent IRRI varieties (Khush and Coffman, 1977). These observations suggest that the yield decline of the highest yielding varieties at each point in time is not caused by viral diseases.

The role of nutrient deficiency or imbalance has received greater attention in recent years. In the LTCCE study, soil samples were taken initially and at periodic intervals thereafter. Soils were analysed for pH, organic C, total N, extractable basic cations, P, and Zn. Unfortunately, all soil samples before 1982 were discarded. Until 1991, nutrient input/output balances including crop removal were not measured, nor were plant tissue nutrient status or components of yield. Zinc has been regularly applied to the LTCCE plots since 1981, but the yield decline has not reversed (Figure 5.8), and soil-test values for potassium and phosphorus have been maintained well above sufficiency levels. Although soil boron (B) is high at IRRI due to relatively high concentrations in the irrigation water pumped from ground water, research on B toxicity has not documented a consistent relationship between plant tissue or soil B status and yield loss under field conditions (Cayton, 1985).

The conservation, and sometimes enrichment of soil organic matter and total N is a consistent feature of long-term experiments in which rice is continuously cropped under submerged soil conditions. In the LTCCE at IRRI where all above-ground biomass is removed at harvest, the concentration of soil organic carbon and total N in the 0–20 cm puddled topsoil has increased slightly since 1963 in control treatments without fertilizer N inputs, and there has been a significant enrichment in soil organic carbon and N in treatments that received high inputs of fertilizer-N (Table 5.5).

Despite this increase, it is impossible to determine whether the total mass of soil organic carbon and N was conserved because measurements of soil bulk density were not made initially, and imposition of the triple rice cropping system with three puddling operation each year may have altered soil physical properties. By 1978, however, 46 consecutive rice crops had been planted and harvested in these plots so that soil physical properties are likely to have stabilized in the period from 1978–1991. In this period, the concentration of soil organic carbon and N remained constant in comparable treatments (Table 5.5), which indicates an equilibrium between inputs and outputs of C and N in this system. Bulk density was measured in the 1991 WS and was similar in all fertilizer N treatments ($0.65/cc^3$, $\pm 0.03 s_{\bar{x}}$) so that the increase in soil organic carbon and N in treatments with greater fertilizer N inputs reflects an actual accumulation of these soil resources compared with the control treatment without fertilizer N inputs. The C/N ratio has remained relatively constant in the 1978–1991 period and has not been influenced by the N input regime.

The homeostasis of C and N in intensively cropped lowland rice soils is remarkable considering the high temperaures of the tropical climate and the removal of all above-ground biomass from the system. This homeostasis results from (i) additional inputs of C from photosynthetic biomass in the floodwater of irrigated rice which is estimated to be 500–1000 kg C/ha per crop cycle (Saito and Watanabe, 1978; Yamagichi et al., 1980; Vaquer, 1984), (ii) inputs of 15–50 kg N/ha per crop cycle from biological $N_2$ fixation by blue-green algae in floodwater and heterotrophic bacteria in soil and the rice rhizosphere (Koyama and App, 1979), and (iii) the slower rate of soil organic matter decomposition by microbes in anaerobic soil compared with rates in aerated soil (Dei and Yamasaki, 1979). Moreover, the microbial populations and biochemical processes that govern the formation of soil organic matter differ greatly in aerated versus anaerobic soil (Hayaishi and Nozaki, 1969; Evans, 1977; Tate, 1979; Benner et al, 1984) so that the chemical nature of organic matter formed under these different environmental conditions may differ. At issue is whether changes in the chemical properties of soil organic matter and the slower rate of organic matter decomposition and N mineralization under anaerobic conditions contribute to the observed yield decline in the long-term experiments due to a reduction in soil N supplying capacity over time.

Soil N availability, plant N status, and N uptake were monitored during the 1991 DS and WS in the LTCCE. Despite 12–15% greater organic carbon and

**Table 5.5** Changes in soil organic carbon and total N of the 0–20 cm topsoil in the Long-term Continuous Cropping Experiment at IRRI. The soil is an Andaqueptic Haplaquoll with a clay texture

| N rate[†] (kg/ha) | | Organic carbon (g/kg) | | | Total N (g/kg) | | | C/N Ratio | | |
|---|---|---|---|---|---|---|---|---|---|---|
| Dry Season | Wet Season | 1963[‡] | 1978 | Mean[§] 1983, 1985, 1991 | 1963[‡] | 1978 | Mean[§] 1983, 1985, 1991 | 1963[‡] | 1978 | Mean[§] 1983, 1985, 1991 |
| 0 | 0 | 18.3 | 18.8 b[¶] | 19.4 b | 1.94 | 1.97 b | 2.00 b | 9.5 | 9.6 a | 9.7 a |
| 50 | 30 | 18.3 | 20.6 a | 20.6 b | 1.94 | 2.13 ab | 2.04 b | 9.5 | 9.6 a | 10.0 a |
| 100 | 60 | 18.3 | 21.4 a | 20.8 b | 1.94 | 2.18 a | 2.10 ab | 9.5 | 9.9 a | 9.9 a |
| 150 | 90 | 18.3 | 21.4 a | 22.3 a | 1.94 | 2.22 a | 2.23 a | 9.5 | 9.7 a | 10.0 a |
| CV% | | 6 | 12 | 8 | 8 | 11 | 7 | 9 | 6 | 4 |

[†] N rate mainplot treatments imposed from 1968–1991.
[‡] Based on soil samples taken from each of six treatments within the four replicate blocks of the experiment. The experimental design and treatments changed in 1968, but blocks remained the same.
[§] Soil samples from 1983, 1985 and 1991 were sampled in the same manner, by the same agronomist, and were kept for future reference. These soils were analyzed in 1992 by the same lab technician, and N-rate treatment means were compared in a repeated-measure AOV with years as subplots.
[¶] Means within columns followed by different letters indicate a significant difference by DMRT, $p < 0.05$.

N content of the high N input treatment compared with the control (Table 5.5), no differences in soil N supplying capacity were evident. Extractable soil $NH_4$-N was similar in all fertilizer N treatments when measured at transplanting before N fertilizer was applied, at panicle initiation just before application of the last N top-dressing, and at the flowering stage (data not shown). Likewise, uptake of N from flowering to physiological maturity was similar in all N fertilizer treatments which indicates no difference in the native N supply from soil resources during this period (Table 5.6). Also consistent with the absence of differences in soil N supply despite differences in soil organic matter content was the lack of significant differences in flag leaf N concentration at the 50% flowering stage, which was comparable and marginally deficient in all N-rate treatments (data not shown).

To test the hypothesis that yields were limited by an insufficient N supply in the LTCCE, N rates were increased in the 1992 DS and varietal subplots were eliminated so that only IR72 was planted in the N-rate main plots. Without fertilizer N inputs, yield of IR72 in the 1992 DS was similar to the yield in the previous three years (1989–1991) in which IR72 was the highest yielding variety (Figure 5.9). Actual N uptake of 50 kg N/ha in the 1-1N plots was equivalent in 1991 and 1992. A yield of 8.7 t/ha was obtained in 1992 with total input of 190 kg N/ha, compared with an indicated yield plateau of 6.3 t/ha from 1989–1991 with input of 150 kg N/ha. Little of the difference in maximum yield attained in the 1991 and 1992 DS could be attributed to differences in solar radiation or temperature regime based on the yield potential predicted by the ORYZA1 simulation model (Kropff et al., 1993). Instead, the greater response to applied N in 1992 versus 1991 was the result of increased N uptake efficiency from applied N (data not shown) and increased rates of N application so that agronomic N use efficiency was greater in 1992, even with the higher rates of N addition (Table 5.7).

**Table 5.6** Nitrogen accumulation during the grain-filling period in the Long-term Continuous Cropping Experiment at IRRI

| N Input Treatment (kg N/ha) | | 1991 DS[†] | | |
|---|---|---|---|---|
| | | Total N/ha (kg N/ha) | | $\Delta$N Uptake (kg N/ha) |
| DS | WS | Flowering | Physiol. Maturity | |
| 0 | 0 | 49 a[‡] | 55 a | 6 |
| 50 | 30 | 69 b | 76 b | 7 |
| 100 | 60 | 88 c | 93 c | 5 |
| 150 | 90 | 107 d | 112 d | 5 |

[†]Values represent the mean of four cultivars at each N rate.
[‡]Mean within columns followed by a different letter differ significantly by DMRT, $p < 0.05$.

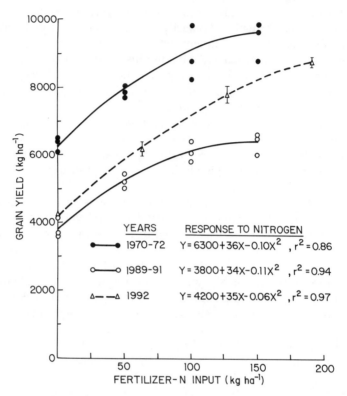

**Figure 5.9** Dry season yield response to fertilizer N rates in the Long-Term Continuous Cropping Experiment at IRRI based on the highest yielding variety each year in three time periods.

We believe that greater uptake efficiency from fertilizer N in 1992 resulted from two factors. First, greater precision of water management and timing of N application at panicle initiation was possible when only one variety was used in 1992. By contrast, the six varieties used as subplots from 1989–1991 had a mean maturity of 112 days from sowing with a range of 100–125 days. Such a wide range in maturity makes it more difficult to time N applications and set the optimal floodwater depth for maximum N fertilizer efficiency (De Datta, 1986). Second, fungicide applications were made to reduce the incidence of sheath blight and stem rot diseases which thrive in the N-rich, high leaf area canopy that is required to achieve a yield of 9 t/ha. Before the 1992 DS, fungicide was never applied in the LTCCE. Although the incidence of sheath blight and stem rot was not monitored before 1992, plant infection by these persistent pathogens was apparent in 1991. With fungicide application in the 1992 DS, there was no sheath blight infection in the three lowest N-rate treatments. In the highest N-rate treatment which received 190 kg N/ha, 9% of the hills were infected by sheath blight and separate

**Table 5.7** Yield response and partial factor productivity for N fertilizer inputs in three periods of the Long-term Continuous Cropping Experiment at IRRI

| Period (cultivar) | Days to Maturity (d) | N Fertilizer rate (kg/ha) | Grain yield (kg/ha) | Yield increase ($\Delta$ kg/ha) | Agronomic N fertilizer use efficiency ($\Delta$ kg grain/kg N) | N Fertilizer factor productivity (kg grain/kg N) |
|---|---|---|---|---|---|---|
| 1970–70 (IR8, IR24) | 127 | 0 | 6300 | — | — | — |
| | | 50 | 7870 | 1570 | 31.4 | 157 |
| | | 100 | 8870 | 2570 | 25.7 | 89 |
| | | 150 | 9430 | 3130 | 20.9 | 63 |
| 1989–91 (IR72) | 115 | 0 | 3800 | — | — | — |
| | | 50 | 5210 | 1410 | 28.2 | 104 |
| | | 100 | 6100 | 2300 | 23.0 | 61 |
| | | 150 | 6380 | 2580 | 17.2 | 43 |
| 1992 (IR72) | 115 | 0 | 4250 | — | — | — |
| | | 63 | 6170 | 1920 | 30.3 | 97 |
| | | 127 | 7760 | 3510 | 27.7 | 61 |
| | | 190 | 8710 | 4460 | 23.5 | 46 |

yield measurements on diseased on healthy hills indicated a 3% yield reduction from this pathogen. These results suggest that sheath blight infection in previous years without fungicide control was a likely constraint to yields in the highest N-rate treatment of the LTCCE. In the early years, sheath blight pressure may have been less than in later years because soil inoculum levels of this pathogen build up over time in continuous, high input rice systems (Mew, 1991).

Partial factor productivity for N fertilizer was similar in 1992 and the 1989–1991 period, but in both periods partial N productivity was substantially lower than in 1970–1972 because the yield without N inputs had decreased from 6.3 t/ha to about 4.0 t/ha in the 1989–1992 period (Table 5.7). In fact, there has not been a decline in the achievable yield potential in the LTCCE, but rather a decline in TFP because crop management has been held constant except for the N fertilizer rates applied as treatments so that a shift in the yield response to applied N fertilizer represents a change in TFP.

It is also clear that a decrease in yield potential of more recent IRRI varieties is not a factor in the decline in TFP of the LTCCE: when yields are corrected for growth duration, there is no difference in maximum attainable yield of IR72 and IR8 or IR24 as calculated from the data in Table 5.7. Likewise, there has not been a change in agronomic fertilizer N use efficiency, which was slightly higher in 1992 than in the 1970–1972 period. Because grain yield is closely related to N uptake by irrigated rice (Cassman et al., 1993), similar agronomic fertilizer N use efficiency indicates that N uptake efficiency by the root systems of early and more recent IRRI varieties has remained constant and is therefore not a contributor to the decrease in N factor productivity. Instead, the major change in the performance of this continuously cropped, intensive rice system is the decrease in yield without N inputs, a decrease which appears to result from a reduction in the N supplying capacity of soil despite the conservation of soil organic matter and total N. This decrease in soil N supply causes a shift in the yield response to applied N fertilizer so that greater N inputs were required to achieve the same yield levels in 1992 than in 1970–72 (Figure 5.9). The farm monitoring data from two intensive rice production domains in the Philippines also indicate a similar shift in the response to applied N such that greater N inputs were required in the 1986–90 period than in 1979–82 (Figure 5.4).

Research at IRRI is now in progress to test two hypotheses concerning the factors that govern the N supplying capacity of soil that is continuously cropped under submerged conditions:

1. that the rate of net N mineralization in anaerobic soil is not substrate-limited so that the soil N supplying capacity is not proportional to the quantity of soil organic matter or total soil N due to the chemical environment of reduced soils and the low energy yield from anaerobic decomposition of organic substrates;
2. that the chemical properties of organic matter fractions formed under anaerobic conditions differ from those of organic matter formed in aerated soil such

that soil humus produced in submerged soils is more recalcitrant to microbial degradation.

Both hypotheses are consistent with the results from the LTCCE where the effective N supply from soil decreases without a concomitant decrease in soil N reserves or a change in the C/N ratio of soil organic matter.

We are hindered in this effort by a lack of data and archived soils from the earlier periods of the long-term experiments. In fact, the IRRI long-term studies have suffered from the perception that these experiments are 'Trials' or 'Long-Term Demonstrations'. Such titles minimize the value and recognition that these experiments warrant, and they do not imply rigorous science and careful measurement. The potential importance of the yield decline phenomenon to rice production highlights the need for quantitative monitoring of the key inputs and outputs that govern productivity in long-term experiments, and the need to monitor changes in soil properties and archive soil samples for future reference.

## 5.7 LINKING TFP TRENDS AT THE EXPERIMENT, FARM, AND REGIONAL LEVELS

More than 90% of all rice is produced and consumed in Asia, and it is estimated that global rice production must increase by 50–70% in the next 30 years. Most of this increase must come from existing intensive, irrigated rice systems in Asia because there is little scope for an expansion of the irrigated area. It is therefore crucial to assess present trends in yield growth at national, farm, and experimental levels as a basis for targeting research and investment for rice-based agriculture. The evidence we have examined does not provide an optimistic scenario. Particularly disturbing is the complete lack of long-term experiments on irrigated rice systems which document that it is even possible to produce sustainable increases in rice production over time.

Yield growth at regional levels, total factor productivity at national, district and farm levels, and partial factor productivity from nutrient inputs in long-term experiments are decreasing where rice is cultivated continuously with irrigation. While there are many factors in both the socio-economic and biophysical environment that may contribute to such trends, a key issue is whether similar underpinning processes are responsible for the decrease in the productive capacity of the paddy environment as irrigated rice systems intensify. Based on the district, farm, and experimental plot data we have examined, it appears that declining partial factor productivity for fertilizer N, and possibly fertilizer nutrients more generally, provide a focal point that deserves closer scrutiny. Monitoring both partial factor productivity and the output/input efficiency from applied nutrients provides a yardstick by which the causes can be identified, their effects quantified, and remedial strategies tested. Moreover, maximizing the efficiency of N use from both endogenous soil reserves and from applied fertilizer N is crucial in sustaining yield growth and increasing profit margins from irrigated rice farming.

The similarity of intercept shifts in the farm-level N response production functions (Figure 5.4) and the N response functions from the LTCCE suggest that the soil resource base has degraded in both cases. To address the issue of whether similar processes are involved, time-trend data on the relationship between soil properties, factor productivity, and crop performance are required. Our work plan at IRRI calls for agronomists to begin monitoring crop performance and biophysical properties of soil in farmers' fields that are included in long-term economic monitoring studies in Central Luzon and Laguna provinces conducted by the Social Sciences Division. Coupled with the ongoing research on soil N supplying capacity in relation to organic matter quantity and quality, other potential constraints such as the sheath blight/stem rot disease complex and off-farm constraints dealing with irrigation infrastructure and the reliability of water supply, this work is pivotal to understanding the basis for sustainable increases in productivity from the irrigated lowlands of Asia.

We have not considered potential negative externalities associated with intensification of rice systems in our analysis. Of major concern are the effects of pesticide use on the on- and off-farm environment, and on human health. Such externalities must be considered in quantifying the total social factor productivity (TSFP). Pesticide effects on human health and the environment have been reported in recent studies from IRRI (Pingali et al., 1994; Rola and Pingali, 1993). When these costs are explicitly accounted for in the estimation of TSFP, then the declining trends in TFP based on the farm monitoring data and long-term experiments discussed here are more dramatic.

## ACKNOWLEDGEMENTS

We thank Mr. Josue Descalsota and Mr. Joven Alcantara for providing the data from the IRRI long-term experiments conducted in the Philippines, Dr. S. K. De Datta for his vision and commitment to maintaining the IRRI long-term experiments, and Dr. Martin J. Kropff for estimating actual yield potential of the 1991 and 1992 dry season rice crops by the ORYZA1 simulation model.

# 6
# Wheat/Fallow Systems in Semi-arid Regions of the Pacific NW America

**B. Duff, P. E. Rasmussen, and R. W. Smiley**

*Columbia Basin Agricultural Research Center, Pendleton, USA*

## 6.1 INTRODUCTION

Wheat (*Triticum aestivum* L.) is a major agricultural commodity in the Pacific North-west and among its most important export crops. There is considerable interest not only in sustaining, but in enhancing the efficiency of wheat production systems in this region. Wheat area grew rapidly from the late 1800s to the first half of the 20th century. During the last 40 years, however, the source of output growth shifted to enhanced yield. Wheat yields in the Pacific Northwest have increased at a near linear rate of 45 kg/ha yr since the 1930s through continuing development of new varieties with higher nitrogen (N) use efficiency (Rasmussen *et al.*, 1989). Better management of cultural operations, notably control of weeds through use of herbicides, has also contributed to increased yield.

There is growing concern among producers, researchers and commodity traders regarding the limits to growth and sustainability of agriculture (Lynam and Herdt, 1989). Sustainability is generally defined as the continuing ability of soil to sustain crop production without degradation of the soil resource base or contamination of the environment. Excess soil erosion remains a constant threat

---

*Agricultural Sustainability: Economic, Environmental and Statistical Considerations.*
Edited by V. Barnett, R. Payne and R. Steiner © 1995 John Wiley & Sons Ltd

## 86  Wheat/fallow systems in semi-arid NW America

to sustained productivity (Larson *et al.*, 1993; Busacca *et al.*, 1985; Nowak *et al.*, 1985). Related to the sustainability issue is the economic viability of Northwest wheat production systems because of strong international competition for export markets, rising input costs, and the increasing costs of externalities such as soil erosion and pesticide use.

When used wisely, long-term experiments provide information on the long-term sustainability of agricultural systems that can be obtained in no other way (Jenkinson, 1991). Oregon State University and the US Department of Agriculture maintain five long-term trials at the Columbia Basin Agricultural Research Center in north-east Oregon, USA (Rasmussen *et al.*, 1989). Several trials date from 1931 and are among the oldest research sites in the Pacific Northwest (Mitchell *et al.*, 1991). We selected one long-term experiment to measure sustainability using the total social factor productivity (TSFP) approach developed by Steiner and Herdt in Chapter 1.

### 6.1.1  Location of the research center

The Research Center is located in the Columbia Plateau geographical province between the Cascade and Rocky Mountains. Topography ranges from nearly level to steeply sloping. Soils consist of loess overlying basalt, and are quite young geologically. The climate is semi-arid, but partially influenced by maritime winds from the Pacific Ocean. Winters are cool and wet, and summers hot and dry. Precipitation occurs primarily during the winter, in contrast to climatic patterns in the midwest and eastern USA. Winter precipitation falls mainly as rain, with limited duration of snow cover in most years. The region is best suited for winter cereal production because of the lack of adequate precipitation from June to August. Water erosion is substantial because of the sloping topography and lack of adequate crop cover during winter months. Rain on snow on frozen soil can occasionally cause severe erosion events.

The Research Center is located on a nearly-level plateau at 455 m above sea level. Average annual precipitation is 420 mm, 70% of which occurs between 1 September and 31 March. The average temperature is 10.2°C, but ranges from −0.6°C in January to 21.2°C in July. Soils are coarse-silty mixed mesic Typic Haploxerolls (Walla Walla silt loam). The upper 30 cm of soil contains 18% clay, 70% silt, and 12% sand.

## 6.2  HISTORY OF THE RESIDUE MANAGEMENT EXPERIMENT

The present report centres on a residue management study initiated in 1931 in response to concerns for 'lower yields, increased production costs, more washing and gullying of the land, and a reduction in the protein content of grain' (Oveson and Besse, 1967). Virgin sod was broken for cultivation in the mid 1880s. The land was cropped annually for many years, but the rotation changed to grain and fallow as yield declined. Fallowing generally enhances or stabilizes yield in semi-arid environments, but soils are sensitive to soil organic matter decline because no crop is being grown over 15 months in a two-year period.

The objective of the experiment was to evaluate the long-term effects of fertilizer amendments and residue management on grain yield and soil quality in a winter wheat/fallow system. Wheat is grown in alternate years, with the fallow year used to store additional soil moisture. The trial has been conducted relatively unchanged since 1931, with modifications introduced only to ensure relevance to modern agriculture. Supporting data from the Center's variety breeding trials are included to evaluate yield improvements over time.

## 6.2.1 Experimental design

The residue management experiment consists of nine treatments (Table 6.1). It has two replications and is divided into two identical series offset by one year so one is in crop while the other is in fallow. Individual plot size is 11.6 by 40.2 m. Winter wheat is seeded in early October (autumn, or 'fall') and harvested in mid-July (summer). Wheat stubble is left undisturbed over the winter, except for a fall-burn treatment. Organic amendments are applied in late March or early April and stubble burned on spring-burn plots. The entire experiment is then mould-board ploughed 20 cm deep and smoothed with a cultivator and harrow. The experiment is then tilled three to four times between April and October to control weeds and maintain seed-zone moisture.

## 6.2.2 Fertilizer amendments

The two organic residue treatments consist of pea vines and barnyard manure applied just prior to ploughing in the spring of the fallow year. These additions have remained essentially unchanged over the 60-year period, with one exception. Pea vine material from 1931–50 included seed with the pods and vines; additions since then have not included the seed. This change significantly lowered the annual N input, with only minor influence on added carbon (C). Partially dried strawy beef manure for the manure treatment has been obtained from the same feedlot since 1931. The dry matter, C, and N content of pea vine and manure additions have been determined since 1976.

Inorganic-N fertilizer has been applied to specific treatments since 1931 as granular materials broadcast and incorporated with tillage. Nitrogen was applied as $NaNO_3$ from 1931–35, as $Ca(NO_3)_2$ from 1936–43, as $(NH_4)_2SO_2$ from 1944–66, and as $NH_4NO_3$ from 1967–92. From 1931–43, N was applied in April of the fallow year. Since then N has been applied in October just prior to planting winter wheat.

**Table 6.1** Straw management, organic residue addition, and inorganic-N fertilizer applied to the long-term residue management experiment between 1932 and 1986, adapted from Rasmussen and Parton (1994)

| No. | Treatment Designation | Straw Management[†] | | | Organic Residue Addition (kg/ha/crop) | | | Inorganic-N Fertilizer (kg/ha/crop) | | |
|---|---|---|---|---|---|---|---|---|---|---|
| | | 1931–66 | 1967–78 | 1979–86 | 1931–66 | 1967–78 | 1979–86 | 1931–66 | 1967–78 | 1979–86 |
| 6 | FB-$N_0$ | FB | FB | FB | 0 | 0 | 0 | 0 | 0 | 0 |
| 7 | SB-$N_0$ | SB | SB | SB | 0 | 0 | 0 | 0 | 0 | 0 |
| 0 | UB-$N_0$ | UB | UB | UB | 0 | 0 | 0 | 0 | 0 | 0 |
| 2 | SB-$N_{45}$ | FD | UB | SB | 0 | 0 | 0 | 0 | 45 | 45 |
| 3 | SB-$N_{90}$ | SD | UB | SB | 0 | 0 | 0 | 0 | 90 | 90 |
| 4 | UB-$N_{45}$ | FD | UB | UB | 0 | 0 | 0 | 34 | 45 | 45 |
| 5 | UB-$N_{90}$ | SD | UB | UB | 0 | 0 | 0 | 34 | 90 | 90 |
| 9 | UB-PV | UB | UB | UB | 2.24 | 2.24 | 2.24 | 0 | 0 | 0 |
| 8 | UB-MN | UB | UB | UB | 22.40[‡] | 22.40 | 22.40 | 0 | 0 | 0 |

[†]Management prior to ploughing in the spring: FB = fall-burn, SB = spring-burn, UB = not burned, FD = fall-disked, SD = spring-disked, PV = pea vines added, MN = manure added.
[‡]No manure applied from 1943 to 1947.

### 6.2.3 Residue burning

Wheat stubble is not disturbed between harvest and burning. The fall-burn treatment is burned in mid-September, and the spring-burn treatments in late March or early April. Burns are implemented by tilling a 1 m border around each plot, starting a backfire on the lee side, and then igniting the windward side. Burning is rapid, with temperatures reaching 300°C in the canopy but rarely persisting for more than three minutes (Rasmussen et al., 1986). The soil surface is not disturbed between burning and ploughing, about 195 and 5 days for fall- and spring-burning, respectively.

### 6.2.4 Adjustments and modifications

Two major changes in management have occurred since 1931. The initial experiment utilized a medium-tall soft white winter wheat variety (cv Rex M-1) and a low rate of N application (34 kg/ha). The experiment was revised in 1967 to change the wheat type from a medium-tall to a semi-dwarf variety, and to expand from one N application rate (34 kg/ha) to two (45 and 90 kg/ha), with duplicate plots of each. This change accommodated the adoption of semi-dwarf wheats with higher yield potential. Wheat cultivars were Nugaines from 1967–73, Hyslop from 1974–78, and Stephens from 1979–92. In 1979, wheat stubble management on one set of the 45 and 90 kg/ha N treatments was changed from non-burn to spring-burn. This change provided a nested factorial of three N-rates and two burn conditions to determine if N fertilization would alter any effects of residue burning.

### 6.2.5 Yield

Grain yield has been determined every year by combine harvesting a portion of each plot. The sample area from 1931 to about 1950 was 10.1 by 40.2 m. Yield between 1951 and 1966 was obtained by harvesting two 3.6 by 40.2 m swaths. From 1967 to 1980, the harvest sample was one 2.1 by 40.2 m swath. Since 1981, yield has been obtained by combining two 2.1 by 20.1 m swaths arranged end to end across the plot.

### 6.2.6 Soil analysis

Soil samples to a depth of 60 cm were taken in 1931, 1941, and 1951 in 30 cm increments, and in 1964, 1976, and 1986 in 15 cm increments. Early soil sampling density is not known. Samples since 1964 consisted of 8–16 cores from the inner portion of each plot. Total N in the 1931–64 samples was determined by standard AOAC macro-Kjeldahl procedures existing at the time of the analysis. Total N in 1976 was determined by the Bremner (1965) procedure after Kjeldahl digestion as outlined by Nelson and Sommers (1972). Total N in 1986 was determined by automated analysis (Technicon, 1976) after digestion as in 1976. Analytical accuracy (coefficient of variation) was similar for all sample times, ranging from 4.7 to 6.8%. Inorganic-N constituted < 1.0% of the total in 1976 and 1986, and there is little reason to suspect it was higher in earlier years. Consequently, N will be referred to as organic-N rather than total N.

Organic matter determinations were made on the 1931 and 1941 soil samples by the loss-on-ignition method of Rather (1917). Because this procedure had a high standard error relative to the last two dates, individual organic-C values were computed by multiplying the mean C/N value by individual N values. An OM/C ratio of 1.72 was used when converting one value to the other. This reduced the standard error for C determination to levels similar to those for combustion analysis. No method of C analysis was reported for the 1951 samples. The 1964 samples were apparently not analysed for C, consequently values for these dates were estimated from C content in 1951 and 1976 and N data for 1964. Organic-C in 1976 and 1986 was determined by dry combustion using a Leco C analyser (Tabatabai and Bremner, 1970).

## 6.3 ANALYSIS AND INTERPRETATION OF RESULTS

### 6.3.1 Yield history

Wheat yields at the Research Center have risen steadily at a linear rate of 52 kg/ha/yr from 2.8 t/ha in 1932 to 5.5 t/ha in 1992 (Figure 6.1). About 45 kg of the increase has been due to development of improved N-efficient, disease resistant varieties and the remainder to improved crop management (weed control, water storage, and water use efficiency). In the residue management experiment, the manure treatment has consistently produced the highest grain and straw yield since its inception (Table 6.2). Grain yield for the manure treatment, which receives adequate N (111 kg/ha/crop), has risen steadily from 2.96 t/ha in the 1930s to 4.87 t/ha in the 1980s. The low fertility treatments (# 0, 6, 7) originally yielded within 20% of the manure treatment, but this percentage has fallen progressively over time until their current yield is 43–57% less. Pea vine treatment yields are very close to those expected from the level of N input (34 kg/ha/crop). The 90 kg N/ha treatment currently yields about 5% less than the manure treatment, but a direct comparison is not possible since manure supplies more N (111 vs 90 kg/ha) and also other elements (P, S, K, Zn, etc).

Both fall- and spring-burning originally tended to increase grain yield, but the effect was short-lived. By the 1950s, yields were similar to those of non-burn treatments, and in recent years, the fall-burn treatment has yielded less than the unburned control. The early yield advantage from burning may have been due to

**Table 6.2** Average grain and straw yields for five time periods between 1932 and 1986, adapted from Rasmussen and Parton (1994)

| No. | Treatment Designation[†] | Period 1932–41 | 1942–51 | 1952–66 | 1967–76 | 1977–86 |
|---|---|---|---|---|---|---|
| | | \multicolumn{5}{c}{Grain yield[‡] (t/ha/crop)} | | | | |

| No. | Treatment Designation[†] | 1932–41 | 1942–51 | 1952–66 | 1967–76 | 1977–86 |
|---|---|---|---|---|---|---|
| 6 | FB-$N_0$ | 2.41 | 2.44 | 1.92 | 2.74 | 2.58 |
| 7 | SB-$N_0$ | 2.48 | 2.56 | 2.08 | 2.83 | 2.67 |
| 0 | UB-$N_0$ | 2.35 | 2.48 | 2.01 | 2.85 | 2.78 |
| 2 | SB-$N_{45}$ | 2.23 | 2.27 | 1.92 | 3.61 | 3.92 |
| 3 | SB-$N_{90}$ | 2.31 | 2.38 | 1.91 | 3.96 | 4.50 |
| 4 | UB-$N_{45}$ | 2.66 | 2.86 | 2.58 | 3.75 | 4.10 |
| 5 | UB-$N_{90}$ | 2.59 | 2.83 | 2.53 | 4.17 | 4.64 |
| 9 | UB-PV | 2.73 | 2.98 | 2.67 | 3.69 | 3.81 |
| 8 | UB-MN | 2.96 | 3.33 | 3.09 | 4.44 | 4.87 |
| | Mean | 2.52 | 2.68 | 2.30 | 3.56 | 3.76 |
| | Srd Error | 0.048 | 0.046 | 0.055 | 0.117 | 0.052 |

Straw yield (t/ha/crop)

| No. | Treatment | 1932–41 | 1942–51 | 1952–66 | 1967–76 | 1977–86 |
|---|---|---|---|---|---|---|
| 6 | FB-$N_0$ | 4.45 | 4.51 | 3.55 | 5.05 | 5.11 |
| 7 | SB-$N_0$ | 4.61 | 4.74 | 3.85 | 5.14 | 5.29 |
| 0 | UB-$N_0$ | 4.33 | 4.60 | 3.80 | 5.00 | 5.21 |
| 2 | SB-$N_{45}$ | 4.26 | 4.20 | 3.56 | 5.96 | 7.04 |
| 3 | SB-$N_{90}$ | 4.30 | 4.41 | 3.53 | 6.32 | 7.59 |
| 4 | UB-$N_{45}$ | 5.18 | 5.57 | 5.04 | 6.03 | 7.03 |
| 5 | UB-$N_{90}$ | 5.06 | 5.51 | 4.93 | 6.37 | 7.46 |
| 9 | UB-PV | 5.39 | 5.91 | 5.19 | 6.21 | 6.95 |
| 8 | UB-MN | 6.40 | 7.22 | 6.71 | 6.78 | 7.94 |
| | Mean | 4.89 | 5.18 | 4.47 | 5.87 | 6.65 |
| | Std Error | 0.100 | 0.080 | 0.073 | 0.211 | 0.121 |

[†]Designation codes are presented in Table 6.1.
[‡]Yield on dry weight basis.

better weed and disease control, or to increased available N because of less N immobilization by residue. But, after 20 years, available N in burn plots had been reduced through organic matter decline to the extent that lower N supply dominated any beneficial effects. Spring-burning has thus far not proven detrimental to production, although it appears to be reducing soil biological activity.

There was a significant increase in grain yield in 1967 with the change from a medium-tall to a semi-dwarf variety. The zero-N treatments exhibited about a 40% increase in yield with little change in available N supply. This was due to both higher transport efficiency of N from vegetative to reproductive tissue and to lower N content of the grain of semi-dwarf wheat.

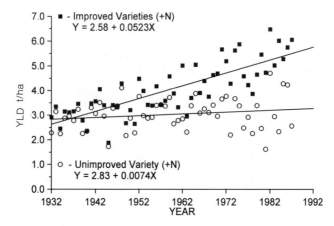

**Figure 6.1** Wheat grain yield change with time, 1932–92. Pendleton, Oregon

### 6.3.2 Soil carbon and nitrogen

With the exception of the manure treatment, soil organic-C and N declined with time for all treatments (Table 6.3). Previous declines have been linear and independent of the initial C and N levels (Rasmussen et al. 1980) This trend continues through 1986 and suggests that an equilibrium is not yet near (Rasmussen and Parton, 1994).

The burn treatments have not lost C in proportion to the estimated C volatility. It is likely that C remaining after the burn is less active biologically and has a longer turnover time (Collins et al., 1992). Fall-burning appears to have a greater detrimental effect than spring-burning, which could be the result of (i) greater soil erosion because of the lack of winter cover, (ii) greater C volatilization than estimated, or (iii) substantial overwinter loss of C due to wind and water action. Personal observations suggest the latter is the most likely explanation of the greater C loss for this experiment.

## 6.4 THE ECONOMIC ENVIRONMENT

Calculating a TSFP index requires input and output prices, estimates of production costs and the cost of externalities resulting from soil erosion and use of fertilizers and pesticides. Annual prices and cost estimates which span the period of the experiment provide the ideal basis for calculating TSFP (see Chapter 1). Such an analysis provides an assessment of both the biological and economic sustainability of each treatment.

A major limitation in the application of the TSFP methodology in this study was the lack of a consistent series of production cost estimates. A limited series is

**Table 6.3** Soil organic-C and organic-N for the 0–30 cm soil depth at six sample dates between 1931 and 1986, adapted from Rasmussen and Parton (1994)

| Treatment No. | Year | | | | | | Year | | | | | |
|---|---|---|---|---|---|---|---|---|---|---|---|---|
| | 1931 | 1941 | 1951 | 1964 | 1976 | 1986 | 1931 | 1941 | 1951 | 1964 | 1976 | 1986 |
| | Organic-C (t/ha) | | | | | | Organic-N (t/ha) | | | | | |
| 6 | 48.67 | 46.44 | 42.15 | 42.45 | 40.01 | 37.01 | 3.71 | 3.56 | 3.28 | 3.16 | 3.02 | 2.85 |
| 7 | 48.12 | 45.74 | 42.69 | 43.18 | 41.42 | 38.96 | 3.66 | 3.51 | 3.32 | 3.33 | 3.20 | 3.06 |
| 0 | 50.21 | 49.23 | 44.93 | 44.00 | 42.47 | 39.65 | 3.83 | 3.78 | 3.50 | 3.43 | 3.29 | 3.19 |
| 2 | 49.24 | 48.08 | 43.78 | 42.72 | 41.31 | 38.74 | 3.75 | 3.69 | 3.41 | 3.41 | 3.27 | 3.20 |
| 3 | 49.41 | 47.08 | 42.80 | 41.81 | 41.52 | 39.95 | 3.77 | 3.61 | 3.33 | 3.35 | 3.33 | 3.30 |
| 4 | 49.90 | 48.36 | 45.96 | 44.91 | 44.24 | 41.36 | 3.81 | 3.71 | 3.58 | 3.62 | 3.48 | 3.46 |
| 5 | 48.75 | 48.59 | 46.18 | 44.09 | 43.65 | 41.90 | 3.71 | 3.72 | 3.59 | 3.46 | 3.41 | 3.35 |
| 9 | 49.92 | 50.48 | 50.63 | 47.01 | 46.15 | 44.49 | 3.81 | 3.87 | 3.94 | 3.77 | 3.66 | 3.58 |
| 8 | 48.57 | 50.88 | 53.07 | 49.39 | 50.80 | 50.18 | 3.73 | 3.90 | 4.13 | 4.08 | 4.18 | 4.20 |
| Mean | 49.20 | 48.32 | 45.80 | 44.39 | 43.51 | 41.36 | 3.75 | 3.70 | 3.56 | 3.51 | 3.42 | 3.35 |
| Std Error | 1.06 | 1.57 | 1.62 | 1.16 | 1.04 | 1.48 | 0.089 | 0.120 | 0.126 | 0.092 | 0.098 | 0.080 |

available for 1974–88 for Oregon (Macnab et al., 1988) and Washington (Painter et al., 1991; Hinman et al., 1981). To remove the limitation, we constructed a proxy for long-term costs based on United States Dept. of Agriculture (USDA) estimates of changes in prices paid by farmers. The first step was to develop a consistent index of input prices for the years 1931–92 (USDA, various years). The index was then normalized with 1992 = 100. The second step was to develop a series of enterprise budgets of the unit costs for each experimental treatment. Because the costs of experimental trials are considerably higher than actual farm costs, the operational and fixed costs for a representative 1,000 ha farm were used as the basis for estimating 1992 production costs for each treatment (Table 6.4). Cultural operations and costs were assumed identical for each treatment. Thus, treatment costs are distinguished only by differences in the costs of N and residue inputs, the cost of yield-dependent operations such as transport, storage, marketing and land use and the level of external costs resulting from pesticide and fertilizer use and soil erosion. Using constant 1992 prices, the cost equation is:

Cost ($/ha) = input costs + yield dependent costs + land use cost
+ externality costs

In step three, individual treatment costs valued at 1992 prices were multiplied by the input cost index from step one to generate annual cost estimates for each treatment. The indexed cost equation is:

Cost ($/ha) = (cost index)(input costs + yield dependent costs
+ externality cost + land use cost).

Because the input price index generated in step one is neither location nor crop specific, we verified the estimates by comparing the calculated indexed costs against a limited survey of actual production costs for the years 1974–88 (Macnab et al., 1988). While the survey was conducted in an area of lower yields than the experimental trials, the results indicate a reasonable correlation ($r^2 = 0.86$) between the two series for the 1974–88 period (Figure 6.2). Differences between the two series are largely the result of higher input levels and inclusion of externalities in the indexed cost series. However, when converted to a cost per ton basis, the indexed costs are lower than the survey costs as a result of higher per-hectare yields.

Cost estimates based on the national index of prices paid by farmers appear to track the movement of the wheat/fallow production costs reported by the survey.

Farm-level wheat prices in Oregon exhibit a moderate degree of variability throughout the period of the experiment (Figure 6.3a), rising slightly over the 1950–72 period followed by a significant rise in 1973 and relative stagnation thereafter. However, both wheat price and yield exhibit a much higher degree of

**Table 6.4** Input costs valued at 1992 prices for selected treatments in crop residue management trials

| Item | Treatment No. 0 1931–92 | Treatment No. 5 1931–66 | Treatment No. 5 1967–92 | Treatment No. 6 1931–92 | Treatment No. 8 1931–92 |
|---|---|---|---|---|---|
| **I. Input costs ($/ha):** | | | | | |
| seed | 17.48 | 17.48 | 17.48 | 17.48 | 17.48 |
| fertilizer: | | | | | |
| inorganic-N | | 18.73 | 49.59 | | |
| organic | | | | | 123.50 |
| herbicides | 15.81 | 15.81 | 15.8 | 15.81 | 15.81 |
| machinery: | | | | | |
| fixed | 120.00 | 120.00 | 120.00 | 120.00 | 120.00 |
| operating | 82.00 | 82.00 | 82.00 | 82.00 | 82.00 |
| crop insurance | 6.45 | 6.45 | 6.45 | 6.45 | 6.45 |
| operating capital (interest) | 18.19 | 20.43 | 24.14 | 18.19 | 33.01 |
| conservation | 0.74 | 0.74 | 0.74 | 0.74 | 0.74 |
| labour | 29.07 | 29.07 | 29.17 | 29.07 | 29.07 |
| total for inputs | $289.74 | $310.72 | $345.28 | $289.74 | $428.06 |

II. Yield-dependant costs:
marketing and handling ($/mt)
field transport  7.35
storage  4.04
elevator handling  4.41
Assessment  0.75
Total  $16.53

III. Land use:
0.35 (share) × wheat price × yield

IV. Externalities:
erosion ($/mt)  $2.73
pesticides

fertilizer application rate for treatment 5 was increased in 1967 with introduction of semi-dwarf varieties.

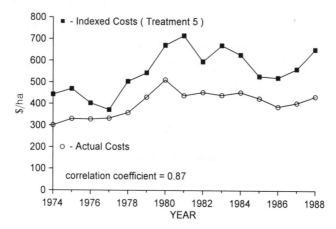

**Figure 6.2** Actual vs indexed dryland wheat production costs in the Columbia Plateau of Oregon and Washington, 1974–88

instability than input costs. The introduction of a succession of government programmes beginning in the 1950s contributed a degree of price stability, although this was disrupted in the late 1970s by the energy crisis and the grain embargo against the Soviet Union. Because much of the wheat produced in the

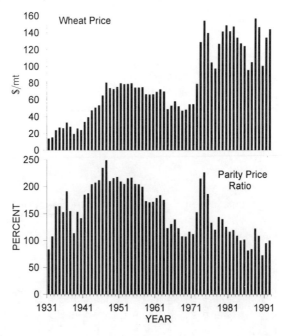

**Figure 6.3** Wheat price (a) and parity price ratio (b) change, 1931–92

Pacific North-west is exported, farm-level prices are affected by conditions in international markets.

Dividing the index of wheat prices by the index of prices paid for production inputs produces a parity price ratio. Long-term wheat prices have increased only slightly since 1970, while input prices have risen steadily. *The result has been a decline in the parity price ratio facing dryland wheat farmers since the mid-1950's* (Figure 6.3b). The parity price ratio does not explicitly embody the effects of technological change and a fall in the ratio does not imply long-term negative returns to farmers provided yields are increasing and/or unit costs are declining. But the parity ratio does identify trends in the relationship between returns and costs.

Externality costs due to on- and off-site effects resulting from pollution and erosion are included in the calculation of TSFP. Soil erosion losses (t/ha/yr) for each treatment were calculated using the Revised Universal Soil Loss Equation (RUSLE) model (Table 6.5). Because the experiment is located on gentle sloping topography, calculated erosion losses are quite low and not representative of losses from steeper slopes found throughout the area. Soil losses were calculated at less than 2 t/ha/yr, in contrast to erosion rates of 10–30 t/ha/yr over much of the Columbia Plateau (Busacca *et al.*, 1985). Calculated soil losses were multiplied by the value estimated by Ribaudo (1989) for the Pacific Region to arrive at an annual cost for soil depreciation and off-farm effects. Accurate assessment of long-term erosion costs is confounded by yield improvements through varietal selection and management, which have masked the long-term impact of erosion on sustainability (Krauss and Allmaras, 1982; Young *et al.*, 1985). To estimate erosion effects on sustainability, we used crop productivity loss equations associated with topsoil erosion developed by Walker and Young (1986) for similar conditions in eastern Washington.

Externality costs associated with fertilizers include regulatory expenses, health effects of nitrates in drinking water, and environmental effects of fertilizer use (such as eutrophication of surface water by phosphorus). Little evidence is available to indicate the extent of nitrate movement into ground and surface water in dryland wheat areas (McCool *et al.*, 1987; Painter and Young, 1993; Painter 1991). However, nitrate in water from some wells in the region, including several near the plots examined in this report, presently exceed the US Environmental Protection Agency guideline for a maximum allowable concentration of 10 mg $NO_3$-N/kg. Drinking water in such areas must be purchased from commercial sources. Other costs associated with nitrate contamination of groundwater in the region are not known.

There are numerous ways that pesticides can enter non-target portions of the environment and cause externality costs and benefits. Costs include regulatory and monitoring expenses, health effects on humans, and environmental effects. The latter include reduced efficiency of the natural enemies of pests, secondary pest outbreaks, pest resistance, crop and tree loss, and poisonings of fish, wildlife, bees, and domestic animals. In contrast, expenses for maintaining roadways, airports,

**Table 6.5** Predicted soil erosion losses (t/ha/yr) for crop residue management trials management trials using the Revised Universal Soil Loss Equation

| Treatment No. | Plot No. | Slope Length (m) | Slope Steepness (%) | 1931 OM (%) | 1931 Residue (kg/ha) | 1986 OM (%) | 1986 Residue (kg/ha) | Predicted soil loss 1931 | Predicted soil loss 1986 |
|---|---|---|---|---|---|---|---|---|---|
| 5 | 05 | 40 | 1.5 | 2.4 | 4563 | 2.0 | 6795 | 0.87 | 0.65 |
|   | 15 | 40 | 3.7 | 2.3 | 4473 | 1.8 | 6527 | 1.88 | 1.68 |
| 6 | 06 | 40 | 1.5 | 2.5 | 1259 | 1.8 | 1520 | 1.03 | 1.05 |
|   | 16 | 40 | 4.5 | 2.2 | 1204 | 1.6 | 1309 | 3.36 | 3.36 |
| 8 | 08 | 40 | 0.9 | 2.5 | 5759 | 2.2 | 7420 | 0.49 | 0.38 |
|   | 18 | 40 | 3.5 | 2.2 | 5679 | 2.2 | 6768 | 1.61 | 1.50 |
| 0 | 10 | 40 | 1.1 | 2.5 | 4071 | 1.9 | 4920 | 0.52 | 0.54 |
|   | 20 | 40 | 1.3 | 2.4 | 3661 | 1.8 | 4393 | 0.69 | 0.67 |

parks, and other public lands are reduced by controlling weeds in cropland. Weed growth or accumulation (wind-blown) along roads clogs drainage culverts and ditches and contributes to roadbed deterioration. Herbicides used in croplands would appear to suppress medical expenses for treatment of allergic reactions to weed pollen. Herbicides improve crop establishment and production of cereal straw as well as grain, which in turn reduces the potential for soil erosion. Herbicides and fungicides reduce the spread of weeds and pathogens to previously uninfested regions, thereby reducing costs for eradication and management in newly infested areas. Fungicides reduce or eliminate hazards associated with contamination of food and feed by aflatoxins derived from fungal growth on unprotected crops.

Pesticides have not been detected in surface or ground water in the dryland wheat production area. It is therefore difficult to assign a net cost or benefit to the pesticide externality, which could be positive rather than negative in dryland cereal production regions. In the absence of further evidence, we simply flagged pesticide and fertilizer externalities in the TSFP calculations by assigning a cost of $2.50/ha.

## 6.5 ESTIMATING TOTAL SOCIAL FACTOR PRODUCTIVITY

### 6.5.1 Definitions

To partition the effect of biological and economic changes in the experiment, two alternative measures of TSFP were estimated. To evaluate *biological sustainability* over time, prices were held constant at 1992 levels. This confines TSFP changes to factors affecting yield. It allows clearer identification of treatment effects during the two time periods. A second procedure was used to measure *economic sustainability*. In this case, both input and output prices were allowed to vary in response to both local and international markets. Economic sustainability is a composite of biological changes and income/cost considerations. Long-term biological sustainability by itself is not a sufficient condition for economic sustainability. Conversely, economic sustainability is neither necessary nor sufficient for biological sustainability.

### 6.5.2 Trends

The effect of varietal development on sustainability from 1932–92 was measured by calculating the mean yield of the three highest yielding varieties included in the Research Center's annual yield trials. In the long run, varietal improvement has had a positive effect on biological sustainability (Figure 6.4). But, because variety trials tend to move to different locations on the Research Center each year, we

## 100  Wheat/fallow systems in semi-arid NW America

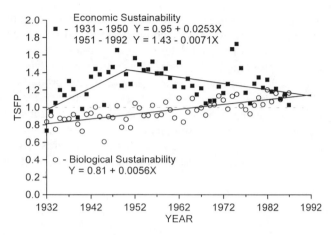

**Figure 6.4** Effect of variety improvement on total social factor productivity (TSFP) using constant (biological sustainability) and indexed prices (economic sustainability)

have no assessment of soil quality change over time. Economic sustainability is sensitive to the overriding impact of the falling output/input price ratio illustrated in Figure 6.3b.

To simplify the presentation, we focused our analysis of TSFP on four of the nine treatments in the long-term residue management experiment. These are treatments #0 (0 kg N/ha, no burn), #5 (90 kg N/ha, no burn), #6 (0 kg N/ha, fall-burn) and #8 (manure, no burn). These treatments encompass most of the effects of residue management on crop yield and soil quality. To examine *biological sustainability* data were divided into two periods: 1931–66, when a single wheat variety was grown, and 1967–92, when improved semi-dwarf wheats were grown. This allows us to isolate the impact of technological improvement on sustainability. To assess *economic sustainability* we examined three time periods: 1931–49, when economic conditions were favourable with little change in technology; 1950–66, when economic conditions were deteriorating with little change in technology; and 1967–92, a period of continuing unfavourable economic conditions, but accompanied by the introduction of improved technology.

Trends in TSFP for treatments #5 and #8 are shown in Figure 6.5 using both constant and indexed prices. Treatment #8 is considered the most stable in the experiment. Biological sustainability (constant prices) did not change during the 1931–66 period, but increased substantially following the introduction of semi-dwarf wheats in 1967. The trend line for treatment #5 showed declining biological sustainability without varietal improvement (1931–66), but a positive trend after adoption of semi-dwarf varieties (1967–92). Both treatments #5 and #8 exhibited positive economic sustainability during the 1931–49 period followed by sharp declines from 1950–66. Improved technology raised the trend-line, but

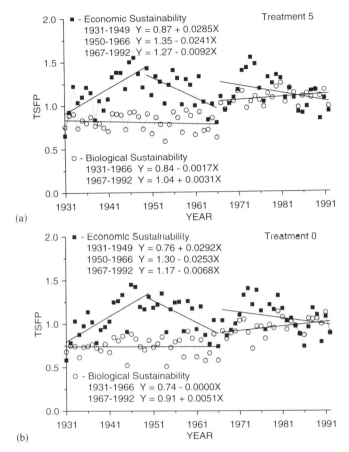

**Figure 6.5** Total social factor productivity (TSFP) for (a) treatment 5 and (b) treatment 8 of the residue management experiment, 1931–92, using constant (biological sustainability) and indexed prices (economic sustainability)

did not arrest the decline in economic sustainability during the 1967–92 period. While wheat farmers systematically improved yields during the 1967–90 period, these gains did not offset the long-term deterioration in price relationships.

A comparison of the sustainability of the low-N input systems (treatment #0, 6) is found in Figure 6.6. Both treatments exhibited declining biological sustainability during the 1931–66 period with #6 declining more rapidly than #0. This is generally due to declining yields associated with decreasing organic-N in the soil. Chemical analysis of the soil shows a continuous decline in organic-N and organic-C from 1931 to 1986 for three of the four treatments (Table 6.3). The slow degradation in the soil resource base is sufficient to offset all increases in productivity during this period. During the 1967–92 period when semi-dwarf varieties were grown, the trend-line is near one, indicating the system is marginally viable. The slope of the trend-line is essentially zero, however, showing that improvements due to varietal selection can do no more than offset the decline brought about the deterioration of the soil resource base. The use of indexed prices, which simulate actual economic conditions, increases the variability of TSFP with an initial increase in economic sustainability during the 1931–49 period followed by declines for both treatments #0 and #6 in the 1950–66 and 1967–92 periods.

The decline in biological and economic sustainability is greatest in treatment #6, in which crop residues are burned and no supplemental N added. Crop residues are required to replenish the organic matter base in semi-arid regions, and removal of this resource trends to accentuate declines in TSFP. Removal of

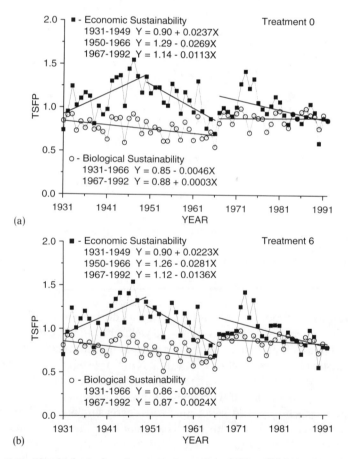

**Figure 6.6** The biological and economic sustainability of (a) treatment 0 and (b) treatment 6 of the residue management experiment, 1931–92

**Table 6.6** Trend estimates of biological and economic sustainability from crop residue management trials, 1931–66 and 1967–92

| Treatment No. | Description | Biological sustainability | | | | Economic sustainability | | | | | |
|---|---|---|---|---|---|---|---|---|---|---|---|
| | | Slope (1931–66) | Rank | Slope (1967–92) | Rank | Slope (1931–49) | Rank | Slope (1950–66) | Rank | Slope (1967–92) | Rank |
| 0 | 0-N, no burn | −0.0046 | 6 | 0.0003 | 7 | 0.0237 | 7 | −0.0269 | 7 | −0.0113 | 7 |
| 2 | 45-N, spring burn | −0.0042 | 5 | 0.0026 | 5 | 0.0232 | 8 | −0.0223 | 1 | −0.0093 | 6 |
| 3 | 90-N, spring burn | −0.0051 | 7 | 0.0031 | 2 | 0.0248 | 5 | −0.0254 | 5 | −0.0090 | 2 |
| 4 | 45-N, no burn | −0.0013 | 3 | 0.0031 | 4 | 0.0278 | 3 | −0.0241 | 3 | −0.0091 | 3 |
| 5 | 90-N, no burn | −0.0017 | 4 | 0.0031 | 3 | 0.0285 | 2 | −0.0241 | 3 | −0.0092 | 4 |
| 6 | 0-N, fall burn | −0.0059 | 9 | −0.0024 | 9 | 0.0223 | 9 | −0.0281 | 8 | −0.0136 | 9 |
| 7 | 0-N, spring burn | −0.0052 | 8 | −0.0016 | 8 | 0.0242 | 6 | −0.0266 | 6 | −0.0129 | 8 |
| 8 | manure, no burn | −0.0004 | 1 | 0.0051 | 1 | 0.0292 | 1 | −0.0253 | 4 | −0.0068 | 1 |
| 9 | pea vine, no burn | −0.0012 | 2 | 0.0015 | 6 | 0.0262 | 4 | −0.0288 | 9 | −0.0092 | 4 |

residues for fuel or likestock feed would produce the same effect as burning. Thus, it is likely that developing countries that routinely utilize all crop residues are more prone to declines in TSFP with time. Treatment #5, which most closely approximates current farm practices, also shows a declining organic-N level although at a slower rate than the zero-N treatment. Organic matter has not declined in the manure treatment (#8). In treatment #5, the yield-augmenting effect of an improved technology has masked some of the continuing decline in organic-N and organic-C in soil. Adoption of semi-dwarf varieties during the 1967–92 period increased the TSFP of all treatments, but had the greatest effect in #5, where inorganic-N input increased to 90 kg N/ha. Treatment #8 responded similarly, reflecting adequate nutrient supply in the manure and no loss of the soil N resource base.

Without the benefit of the continuing development of intensively managed N responsive varieties, all nine treatments exhibited negative coefficients for biological sustainability during the 1931–66 period (Table 6.6). With the adoption of improved technology beginning in 1967, only the zero-N fall and spring-burn treatments continued to decline. All treatments showed increasing economic sustainability from 1931 to 1949. However, because of the long-term decline in the parity price ratio, no treatment was economically sustainable after 1950. Introduction of improved technology in 1967 increased the TSFP for all treatments and reduced, but did not arrest, the rate of decline in economic sustainability.

### 6.5.3 Influence of soil erosion

Organic-C loss from soil for the zero-N treatment (#0) for the 1931–86 period was 10.56 t/ha. Based on average OM content and soil bulk density, this equates to an erosion loss of 15.2 t soil/ha/yr. The calculated soil loss (RUSLE) for this treatment was only 0.6 t/ha/yr, or less than 5% of the total. *Thus, in a wheat/fallow farming system, biological oxidation produces a soil quality loss equal to an erosion loss of 14.6 t/ha/yr.* Biological oxidation of OM is a unique feature of crop rotations that include fallow. It is essentially independent of erosion loss; and created primarily by reduced C input in a wheat/fallow system (no residue is produced in the fallow year).

Therefore, if soils in a wheat/fallow system are allowed to erode at the acceptable USDA soil loss ('T'-value) of 11 t/ha/yr (Nowak *et al.*, 1985), the actual C loss from the system approximates an erosion loss of 26 t/ha/yr. A 2T loss equates to a soil loss of 37 t/ha/yr. The effects of these loss rates on TSFP are illustrated in Figure 6.7. This figure was developed for treatment 5 (90 kg N/ha, no burn), and includes losses in yield potential with continuing soil OM decline. The major effect of erosion is to drop the trend-line substantially, indicating decreasing profitability. There is also a reduction in long-term biological sustainability. It must be cautioned, however, that the long-term loss in sustainability is

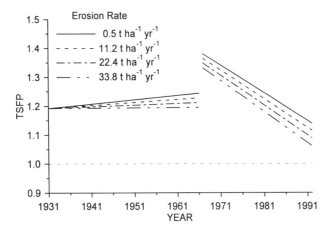

**Figure 6.7** Sensitivity of economic sustainability to alternative soil erosion rates in treatment 5 of the residue management experiment, 1931–92

conservative because soil OM loss becomes progressively more detrimental as topsoil is removed (Young et al., 1985).

### 6.5.4 Prices, profits and government programmes

To examine the sensitivity of sustainability to changing economic conditions, the price of wheat was varied from $91.88/t to $202.13/t in treatment #5, which most closely approximates actual farm conditions. The solutions generated by these alternative prices indicate that price acts to shift the trend-line (Figure 6.8) but in general changes the slope of the line only slightly through its interaction with price dependent costs such as land use.

Total social factor productivity measures the direction but not the level of economic profitability. To quantify profitability, indexed costs are subtracted from revenue. Figure 6.9 illustrates changes in profitability for treatment #5 from 1931 to 1992. Profits rose slowly until the 1960s, jumped dramatically in the 1970s with the introduction of the semi-dwarfs and have fallen nearly continuously throughout the 1980s and early 1990s. These results are symmetrical with those for economic sustainability shown in Figure 6.5.

Using a simplistic approach we can approximate the current effect of government price supports by assuming the 1,000 hectare farm is subject to a $50,000 limitation in program payments. To be in compliance with programme requirements, five % of the wheat base is removed from production, leaving 475 hectares in wheat. Programme participation adds the cost of fallowing 25 idle hectares ($123.50/ha) to production costs producing a net increase in revenue of $98/ha. With a yield of 4 t/ha, this is equivalent to a price increase of $24.50/t. Thus, the

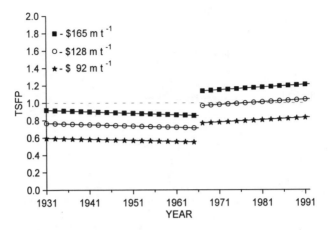

**Figure 6.8** Sensitivity of total social factor productivity to changes in wheat price, treatment 5, residue management experiment, 1931–92

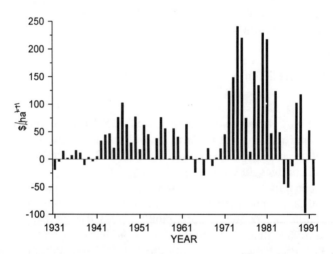

**Figure 6.9** Profitability of treatment #5 (90-N, no burn) with indexed prices (1931–92).

direct effect of programme participation is to shift the sustainability curve upward. However, unless the costs of externalities such as erosion and nitrogen leaching are included, price supports may result in suboptimal cropping patterns and understate their impact on total costs (Painter and Young, 1993; Painter, 1991). Conversely, Federal conservation programmes may reduce erosion, but increase production costs because of their effect on field geometry. While we recognize the crucial role of government in enhancing biological (research, conservation) and economic (price support, export enhancement) sustainability, because of the complexity of these programs, we have focused on a baseline case which assumes no price support. Government programme payments to Oregon wheat farmers from 1987–91 ranged from 8–31% of total value of production for participating farmers (Oregon Dept. Agriculture, 1992).

## 6.6 IMPLICATIONS FOR BIOLOGICAL AND ECONOMIC SUSTAINABILITY OF WHEAT/FALLOW SYSTEMS

Without the benefit of technological improvement (varieties, nitrogen application, intensive management of weeds, and better equipment for cultivating, planting, and harvesting), biological sustainability declined when the soil resource base declined (1931–66). With technological improvement (1967–92), biological sustainability rose. High input treatments show a positive trend over time, whereas low input treatments continue the negative trend. In these treatments, a continuing decline in the resource base is sufficient to overcome improvements achieved through variety selection but no nitrogen input. This indicates that the soil resource base affects long-term sustainability. With high inputs, biological sustainability is profitable and the long-term trend is positive.

Factoring economics into biological sustainability affects long-term trends significantly. Early trends (1931–49) were positive due to increasing yield and a favourable parity price ratio. Trends for later periods (1950–66 and 1967–92) are generally profitable (TSFP > 1) but the trend is negative in all wheat/fallow systems. Costs are continuing to rise while wheat prices remain static. While high input systems remain profitable (TSFP > 1), the trend-line is distinctly negative. If this continues, most systems will lose profitability by the year 2000.

When soil erosion is factored into the question of sustainability of profitability of biological sustainability is substantially reduced, and sustainability trends become more negative. This suggests that excessive erosion such as now occurs cannot be tolerated indefinitely. Both biological and economic sustainability are reduced even when soil erosion occurs at the accepted T-value of the Revised Universal Soil Loss Equation.

## 6.7 SUMMARY

Development of higher yielding varieties which are more water efficient and disease resistant appears unable to overcome the decline in biological sustainability in the semi-arid Pacific North-west wherever there is a decline in the soil resource base. Economic sustainability is presently declining under all wheat/

fallow system practices because production costs are rising faster than wheat price or yield increases.

The wheat/fallow rotation system was designed to reduce the risk of crop failure resulting from inadequate soil moisture. By minimizing evapotranspiration demand during a 15 month non-growing period through clean cultivation or surface mulch techniques, the rotation permits a single cereal crop to be grown every two years using cumulative moisture retained in the soil profile. The system is effective in reducing yield variability resulting from inadequate growing season precipitation and produces higher yields compared with annual cereals grown under identical conditions.

As highlighted by analysis of the long-term trials, a major limitation of the wheat/fallow system is the accelerated loss in soil organic matter which accompanies the non-productive part of this rotational pattern. The high oxidation soil rate is not balanced by an equivalent return of organic residues during the cropping portion of the rotation. In cases where residues are removed or burned, the decline in soil organic matter is accelerated. Only in the case in which large amounts of manure were applied was the decline arrested. And while use of both organic and inorganic soil amendments in conjunction with improved cultivars enhances the biological sustainability of the system, increasing input costs combined with static wheat prices makes the system economically unsustainable.

What options are available to arrest the long-term decline in the vitality of dryland soils? Clearly any solution must balance the needs of biological sustainability with the economic well-being of those who derive their livelihood from the health of the soils they farm. The atomistic nature of agriculture and the growing importance of international markets makes it unlikely that farmers individually or collectively will be able to affect product or input prices significantly. Nor is it likely that massive public support for direct income transfers to agriculture through price support and similar programmes will continue indefinitely.

Adoption of conservation tillage on sloping landscapes must be encouraged or supported financially. A complementary option would be to divert wheat/fallow systems periodically to perennial grass (perhaps 10 of 40 years) to maintain the soil resource base. Strong emphasis should also be placed on adequate erosion control, since soil loss increases organic matter loss and decreases total social factor productivity. Erosion coupled with oxidation losses produced negative long-term biological and economic sustainability in practically all instances.

Perhaps the most viable option lies in development of technologies and management systems which reduce production costs while maintaining or increasing yields, and using land, water and other system resources more efficiently. To improve biological sustainability, factors such as soil erosion and oxidation which deplete soil productivity must be reduced. Annual cropping rotations using cultivars with reduced transpiration requirements could reduce and reverse the decline in soil organic matter without increasing the risk of crop failure. The ability to grow crops with less fertilizer, pest- and disease-controlling chemicals and to use machinery more efficiently might lower costs and increase economic sustainability. An added benefit of these developments would be a reduction in environmental and other externality costs.

Such solutions are not likely to come from industry nor from farmers themselves, but from a continuing partnership which agricultural research plays a prominent role.

# 7
# Multi-crop Comparisons on Sanborn Fields, Missouri, USA

J. R. Brown, D. D. Osburn, D. Redhage and C. J. Gantzer

*University of Missouri, USA*

## 7.1 INTRODUCTION

In 1888, J. W. Sanborn initiated a study to demonstrate techniques that farmers might use to offset the decline in soil productivity in Missouri (Anon., 1988). His study, initially called the Rotation Field, involved the use of farmyard manures and different crop sequences or rotations (Upchurch, *et al.*, 1985).

The Rotation Field, renamed Sanborn Field in 1924, has been in continuous culture since 1888. Although periodic changes have reflected the changes in agricultural production over time, the focus on rotations has continued. The value of crop rotation and manure was demonstrated early and this permitted application of newer techniques to sustain productivity. In spite of other changes over time, the treatments on a core of plots have not changed since 1888.

The efforts on Sanborn Field and elsewhere since 1888 have shown that under a given cropping and fertility system there are long-term changes to the soil resource. In fact, soil fertility at some locations has increased to a point where people are concerned about the pollution of ground-water from excess nitrogen (N) or of surface waters from excess phosphorus (P). In addition, crop production practices that may add metals such as cadmium or persistent toxic organic compounds to the soil are causing considerable concern.

These expressions of concern over the past two decades have led to increased attention to the sustainability of agriculture. Inherent in this movement toward

---

*Agricultural Sustainability: Economic, Environmental and Statistical Considerations.*
Edited by V. Barnett, R. Payne and R. Steiner © 1995 John Wiley & Sons Ltd

sustainability is the necessity to maintain the productive capacity of the soil resource to sustain humankind in the future. Reducing inputs alone may not maintain the ability of soils to sustain production.

Lyman and Herdt (1989) defined sustainability as 'the capacity of a system to maintain output at a level approximately equal to, or greater than, its historical average, with the approximation determined by its historical level of variability'. Inherent in a sustainable system must be the maintenance of the soil productive capacity.

The objective of this study was to use the input and output data obtained from Sanborn Field under different crop sequences to determine which management schemes were sustainable according to the Lyman–Herdt definition.

## 7.2 SELECTION OF STUDY MATERIALS

Changes were made to several plots in 1914, 1928, 1940, and/or 1950 (Upchurch, et al., 1985). There were nine plots which continued until 1989 without alteration. However, as Wagner (1989) explained, each time the management system is changed, the soil tends toward a new equilibrium level which may take 20–40 years to achieve. Also, residual effects of past treatments persist for several years (Motavalli, et al., 1992). Therefore, if sustainability indices of crop production are to be compared, the period of the study should be of sufficient length to allow some semblance of equilibrium to be reached in the soil parameters.

Based upon length of continuous culture, only two periods in the history of Sanborn Field would qualify for sustainability comparisons (i.e., 1888–1913 [26 crop years] and 1950–89 [40 crop years]). The 1950–89 period was selected for this study because the practices and crop sequences more nearly reflect the practices used by modern farmers. In most cases, the plots selected for study had been in a similar management programme prior to 1950 and therefore new equilibrium levels were probably achieved early.

## 7.3 DESCRIPTION OF MANAGEMENT

Eighteen of the 38 plots on Sanborn Field were selected (Table 7.1). The increased availability of fertilizer and the reduction in the demand for on-farm forages due to mechanization after World War II drastically changed the appropriateness of crop production practices. Those practices put into effect in 1950 (Table 7.1) reflected the technology in use at that time.

These management practices were employed on plots $30.5 \times 9.5 \, m^2$ in size each separated by a 1.5 m grass border. The original 1888 design was unreplicated and unrandomized and thus limits the proper statistical evaluation of the data.

**Table 7.1** Summary of the management programme for the Sanborn Field plots selected for sustainability evaluations 1950–89

| Plot No. | Crop Sequence[†] | Treatment[‡] |
|---|---|---|
| 1, 3, 4 | Corn–Wheat–Red clover | Full Fertility |
| 2 | Continuous Wheat | Full Fertility |
| 6 | Continuous Wheat | None |
| 10 | Continuous Wheat | Manure |
| 9 | Continuous Corn | Full Fertility |
| 17 | Continuous Corn | None |
| 18 | Continuous Corn | Manure |
| 25 | Corn–Wheat–Red clover | Manure + 112 kg N/ha Corn + 37 kg N/ha Wheat |
| 26 | Corn–Wheat–Red clover | Full Fertility |
| 27 | Corn–Wheat–Red clover | None |
| 28 | Corn–Wheat–Red clover | Lime, Starter on Corn and Wheat |
| 34 | Corn–Oat–Wheat–Red clover | Manure |
| 35 | Corn–Oat–Wheat–Red clover | None |
| 36 | Corn–Oat–Wheat–Red clover | Full Fertility + Lime |
| 37 | Corn–Oat–Wheat–Red clover | Starter on Corn and Wheat |
| 38 | Corn–Oat–Wheat–Red clover | Lime, Starter on Corn and Wheat |

[†]Corn = *Zea mays* L.; wheat = *Triticum aestivum* L.; red clover = *Trifolium repens* L.; Oat = *Avena sativa* L.
[‡]Full = nutrients applied according to soil test levels and current Missouri extension recommendations.
Manure = 13.4 mg/ha/yr manure (wet basis) since 1888.
None = no treatment since 1888.
Starter = 9-12-8 (kg N-P-K/ha).

The plots on Sanborn Field have been mold-board ploughed in the fall prior to seeding corn and oat in the following spring. Ploughing for winter wheat seeding is done as soon as the preceding crop is removed. Manure treatments have been immediately ploughed under and most fertilizers and limestone have been applied prior to the secondary tillage used to prepare the seedbed. The 'full' fertilizer treatment has been based upon soil-test results and the fertilizer recommendation programme used by farmers. Yield goals, required for some fertilizer recommendations, take into account the production potential of the soil and have been set above the Missouri state average for the given crop.

Seeding has been carried out with mechanical equipment using appropriate seeding rates. Cultivation of corn was done as needed. In recent years, herbicides which have varied as chemical improvements have been made, have been applied prior to corn emergence. However, attempts have been made to use the same variety of each crop for several successive years.

## 7.4 PROCEDURES

The approach taken for the analysis of the 1950–89 results was to assume Sanborn Field was a representative Central Missouri farm. The fact that manure was used implied that animals were reared. However, it is assumed that such manure was available from an adjacent farm and was 'donated' without charge except for spreading and no credit was given for the value of the nutrients contained in the manure. In practice the use of manure declined on farms during the 1950–89 period in Central Missouri due to the gradual shift to cash-crop farming.

### 7.4.1 Fertilisers

Mononutrient carriers were used for fertilizer sources and were appropriately mixed where starter fertilizer was required. Rock phosphate was used on some plots. Several plots received manure on an annual basis but the manure was never analysed to determine its nutrient level and no records were made of its source. Therefore, the amount of N, P, and K added by the manure applications is unknown. Manure was treated as a waste product in this study and no value was assigned to it other than the application cost.

The nutrient application rates followed the 1950 plan which included replacement of nutrients removed by the grain. The N, P, and K concentrations for each of the grains were those given by Morrison (1947) and are summarized in Table 7.2.

### 7.4.2 Soil characteristics

In 1950, at the start of the period used in this sustainability evaluation, soil testing and fertilizer use were in their infancy. The calibration work necessary to validate soil tests for recommendation purposes was just getting underway. Therefore, the initial fertilizer treatments were based on crop removals. By the late 1950s, the calibration work was sufficiently advanced that recommendations were then based upon the level of P, K, and acidity in the soil and not on grain removal. Nitrogen was, and continues to be, recommended based on the yield goal. This yield goal has changed with time as crop yield potentials have been increased due to breeding and selection.

Soil test techniques used in 1949 on samples from Sanborn Field were sufficiently different from those in later years that direct comparisons of results are questionable. However, the test results from 1963 and 1989 were made using very similar methods (Table 7.3).

The soil test levels assumed adequate for optimum production are 22 µg P and 150 µg K per g of soil. The optimum pH is considered to be 6.1–6.5 (Buchholz,

**Table 7.2** Composition of grains (Morrison, 1947)

| Nutrient | Corn | Grain (%) Wheat | Oat |
|---|---|---|---|
| Nitrogen | 1.50 | 2.10 | 1.92 |
| Phosphorus | 0.27 | 0.43 | 0.33 |
| Potassium | 0.31 | 0.44 | 0.40 |

1983). The 'full fertility' treatments were more than adequate to maintain extractable P at the optimum level (Table 7.1, footnote). The full K treatment had mixed results. It is clear that field management failed to maintain optimum pH at 'full treatment'. The full treatments included N applied as ammonium nitrate which, over time, tends to acidify soils.

**Table 7.3** Soil-test results from Sanborn Field samples taken in 1963 and 1985–0 to 20 cm*

| Plot | Treatment | OM (%) 1963 | 1989 | P(µg/g) 1963 | 1989 | K(µg/g) 1963 | 1989 | pH 1963 | 1989 |
|---|---|---|---|---|---|---|---|---|---|
| 1 | Full | 2.2 | 2.4 | 40 | 51 | 122 | 117 | 5.4 | 5.7 |
| 3 | Full | 2.1 | 2.2 | 30 | 44 | 125 | 138 | 5.8 | 6.5 |
| 4 | Full | 2.2 | 2.0 | 40 | 32 | 115 | 132 | 4.9 | 4.5 |
| 2 | Full | 2.2 | 1.9 | 42 | 45 | 129 | 156 | 6.0 | 5.8 |
| 9 | None | 1.2 | 1.0 | 12 | 8 | 136 | 103 | 4.2 | 4.5 |
| 10 | Manure | 2.0 | 2.4 | 60 | 95 | 305 | 395 | 4.4 | 5.8 |
| 6 | Full | 2.2 | 1.9 | 66 | 99 | 230 | 288 | 5.6 | 5.6 |
| 17 | None | 1.2 | 1.0 | 6 | 5 | 141 | 124 | 4.2 | 4.5 |
| 18 | Manure | 2.0 | 2.4 | 32 | 82 | 212 | 294 | 4.6 | 6.2 |
| 25 | Manure +N | 2.4 | 3.2 | 22 | 42 | 200 | 100 | 4.6 | 5.6 |
| 26 | Full | 1.9 | 2.4 | 15 | 40 | 120 | 246 | 6.4 | 5.0 |
| 27 | None | 1.8 | 2.0 | 4 | 3 | 102 | 118 | 4.5 | 5.1 |
| 28 | Lime, P, K Starter | 2.0 | 2.3 | 17 | 32 | 79 | 110 | 6.4 | 5.6 |
| 34 | Manure | 2.5 | 2.4 | 30 | 44 | 228 | 118 | 4.8 | 5.7 |
| 35 | None | 1.8 | 1.9 | 5 | 4 | 140 | 52 | 4.4 | 5.0 |
| 36 | Full | 2.1 | 2.2 | 15 | 25 | 142 | 98 | 5.3 | 6.0 |
| 37 | Starter | 2.0 | 2.0 | 10 | 8 | 128 | 71 | 4.6 | 5.5 |
| 38 | Lime, Starter | 2.1 | 2.2 | 8 | 8 | 95 | 62 | 6.2 | 6.6 |

*Organic matter (OM) by a modified Walkley–Black procedure, P extractable by 0.025 N HCl + 0.03 N NH$_4$F, K extractable by 1 M ammonium acetate @ pH 7.0, pH in 1:1 soil:0.01 M Ca Cl$_2$ (Brown and Rodriguez, 1983).

**Table 7.4** Soil-test results with depth from three plots* in 1989

| Depth Plot (cm) | OM(%) | | | P(µg/g) | | | K(µg/g) | | |
|---|---|---|---|---|---|---|---|---|---|
| | 6 | 9 | 10 | 6 | 9 | 10 | 6 | 9 | 10 |
| 0–10 | 1.9 | 1.4 | 2.7 | 104 | 10 | 95 | 306 | 99 | 386 |
| 10–20 | 2.0 | 1.4 | 2.8 | 96 | 7 | 95 | 260 | 108 | 414 |
| 20–30 | 1.4 | 1.0 | 1.8 | 20 | 3 | 34 | 273 | 144 | 345 |
| 30–40 | 1.0 | 0.9 | 1.3 | 9 | 3 | 6 | 223 | 134 | 226 |
| 40–50 | 0.8 | 0.8 | 1.0 | 8 | 6 | 10 | 182 | 127 | 157 |
| 50–60 | 0.6 | 0.7 | 0.8 | 12 | 9 | 14 | 142 | 114 | 132 |
| 60–70 | 0.6 | 0.6 | 0.6 | 8 | 8 | 14 | 128 | 110 | 135 |
| 70–80 | 0.6 | 0.6 | 0.5 | 8 | 8 | 15 | 122 | 101 | 121 |
| 80–90 | 0.5 | 0.6 | 0.5 | 10 | 10 | 15 | 122 | 90 | 116 |
| 90–100 | 0.4 | 0.5 | 0.5 | 10 | 10 | 15 | 116 | 78 | 106 |

*See Table 7.1 for management information on plots 6, 9 and 10 and Table 7.3 for identification of the soil-test methods.

The manure treatments increased extractable soil P and pH between 1963 and 1989, but had mixed effects on extractable soil K. The low input treatments on plots 28, 37, and 38 failed to maintain extractable P and K levels, clearly demonstrating the importance of nutrient management.

The characteristics of the entire soil profile can have an impact on the ability of the soil to sustain production. The data in Table 7.4 illustrate the differences in three soil properties with depth of three plots within an area of $30 \times 60 \, m^2$ on Sanborn Field. Even though corn was the crop on plot 6 and wheat on 9 and 10, the subsoil fertility indices were much lower on plot 9 (no treatment) than plots 6 and 10 (full treatment and manure, respectively). This occurred because the lack of nutrient additions to the surface soil forced the plants grown on plot 9 to draw nutrients from the subsoil. Thus, another facet of measuring the sustainability of a practice is to determine the effect on the entire rooting depth of the soil.

### 7.4.3 Soil erosion

Crop management affects soil erosion through the effects of tillage, canopy cover, quantity of biomass produced and extent of residue cover when a crop is not growing. The arrangement of Sanborn Field, started more than 30 years before the initiation of the scientific study of erosion, did not permit the installation of erosion measuring devices when they became available.

Core sampling of the individual plots every 25 years permits a rough quantitative estimate of erosion using the depth to the top of the argillic horizon as a reference base. It has become evident that erosion under unfertilized corn was severe because the upper argillic horizon had become part of the tilled horizon.

The Claypan Farm, a facility of the Agricultural Research Service, United States Department of Agriculture, located about 35 km east of Sanborn Field, is a source of erosion data from soil similar to that found on Sanborn Field. The Universal Soil Loss Equation (USLE) (Wischmeier and Smith, 1978), which is based on such data, can be used to estimate soil loss under different cropping practices.

In this study, the costs of erosion per ton of soil were evaluated using a Conservation Reserve Program study by Ribaudo (1989). The USLE is: $A = RKLSCP$ where

$A$ = Soil loss in tons (US) per acre per year
$R$ = Rainfall factor
$K$ = Soil erodibility factor
$LS$ = Slope length and steepness factor
$C$ = Cropping management factor
$P$ = Erosion control practice factor

An $R$-factor of 225, used for all plots, is the average annual Central Missouri value and incorporates the effects of rainfall events. The soil erodibility factor or $K$-factor was determined using a $K$-factor nomograph in USDA Agricultural Handbook 537 (Wischmeier and Smith, 1978). The slope-effect chart in USDA Agricultural Handbook 537 was used to determine the $LS$ or slope length and steepness factor. Eight slope measurements were taken on each plot and averaged to determine the slope.

Slope lengths on Sanborn Field vary depending on the plot. Sanborn Field differs from typical farms in that the slope lengths do not exceed 30 m because of the slope direction of the plots. For plots 1–4 and 6, the slope length is 30 m. For the remaining plots, a 23 m slope length was used because the slopes angled across the plots.

Initial $C$-factor values were those published by Steichen (1979). It was understood from the start of the computations that these values were based on high crop yields. Sanborn Field contains some plots that receive less than full fertility treatments: this tends to delay canopy development and reduces residue production. In fact, on some untreated plots, full canopy coverage is never achieved which increases the $C$-factor. To obtain a more accurate $C$-factor for each plot, an individual $C$-factor was calculated for each crop and management level involved in the study.

An unpublished study carried out by Steiner (1992), based on data from Rodale, included an estimate of the external cost of soil depreciation. The EPIC model (Williams, *et al.*, 1984) was used to evaluate the long-term yield. Any deviation from this yield was attributed to soil depreciation (Steiner, personal communication, 1992). After discussing this with Dr Steiner (Steiner, personal communication 1992), it was decided that insufficient data were available to model soil depreciation properly. The complexities of such a model include the influence of the first 60 years of unfertilized cropping yield changes due to

## Multi-crop comparisons on Sanborn Field

**Table 7.5** Average crop yield by periods from Sanborn Field

| Species | Plot | Management | 1950–1962 | Period 1963–1969 Yield (kg/ha) | 1970–1989 |
|---|---|---|---|---|---|
| corn | 1,3,4 | Rotation, Full Fertility | 4848 | 6941 | 6652 |
| | 6 | Continuous, Full Fertility | 4803 | 6665 | 7511 |
| | 17 | Continuous, None | 640 | 784 | 533 |
| | 18 | Continuous, Manure | 2633 | 3191 | 3749 |
| | 25 | Rotation, Manure Plus | 4767 | 7700 | 7806 |
| | 26 | Rotation, Full Fertility | 5423 | 6151 | 7361 |
| | 27 | Rotation, None | 3731 | 3398 | 3561 |
| | 28 | rotation, Starter | 5166 | 6483 | 5380 |
| | 34 | Rotation, Manure | 4602 | 5543 | 6165 |
| | 35 | Rotation, None | 3423 | 4038 | 4715 |
| | 36 | Rotation, Full Fertility | 4871 | 5022 | 6765 |
| | 37 | Rotation, Starter | 3875 | 4740 | 5712 |
| | 38 | Rotation, Lime, Starter | 4555 | 5392 | 5687 |
| clover | 1,3,4 | Rotation, Full Fertility | $4.48 \times 10^3$ | $9.03 \times 10^3$ | $7.39 \times 10^3$ |
| | 25 | Rotation, Manure Plus | $2.76 \times 10^3$ | $10.15 \times 10^3$ | $9.18 \times 10^3$ |
| | 26 | Rotation, Full Fertility | $3.63 \times 10^3$ | $8.71 \times 10^3$ | $8.58 \times 10^3$ |
| | 27 | Rotation, None | $1.57 \times 10^3$ | $5.62 \times 10^3$ | $4.86 \times 10^3$ |
| | 28 | Rotation, Starter | $4.66 \times 10^3$ | $10.04 \times 10^3$ | $8.31 \times 10^3$ |
| | 34 | Rotation, Manure | $4.86 \times 10^3$ | $10.82 \times 10^3$ | $11.13 \times 10^3$ |
| | 35 | Rotation, None | $3.44 \times 10^3$ | $7.32 \times 10^3$ | $7.19 \times 10^3$ |
| | 36 | Rotation, Full Fertility | $5.24 \times 10^3$ | $8.85 \times 10^3$ | $10.95 \times 10^3$ |
| | 37 | Rotation, Starter | $3.18 \times 10^3$ | $8.78 \times 10^3$ | $9.43 \times 10^3$ |
| | 38 | Rotation, Lime, Starter | $4.75 \times 10^3$ | $8.87 \times 10^3$ | $9.92 \times 10^3$ |
| wheat | 1,3,4 | Rotation, Full Fertility | 2150 | 3273 | 3938 |
| | 2 | Continuous, Full Fertility | 1976 | 3158 | 2675 |
| | 9 | Continuous, None | 430 | 1015 | 638 |
| | 10 | Continuous, Manure | 1673 | 3051 | 2567 |
| | 25 | Rotation, Manure Plus | 1613 | 2943 | 3259 |
| | 26 | Rotation, Full Fertility | 2177 | 3400 | 3515 |
| | 27 | Rotation, None | 612 | 1767 | 2090 |
| | 28 | Rotation, Starter | 1613 | 3071 | 3091 |
| | 34 | Rotation, Manure | 1828 | 3037 | 3837 |
| | 35 | Rotation, None | 1539 | 2083 | 2856 |
| | 36 | Rotation, Full Fertility | 2298 | 3669 | 4509 |
| | 37 | Rotation, Starter | 2036 | 4019 | 3582 |
| | 38 | Rotation, Lime, Starter | 2339 | 3884 | 3662 |
| oat | 34 | Rotation, Manure | 1176 | 2925 | 2806 |
| | 35 | Rotation, None | 1061 | 2774 | 1943 |
| | 36 | Rotation, Full Fertility | 1692 | 2867 | 3000 |
| | 37 | Rotation, Starter | 1330 | 2978 | 2408 |
| | 38 | Rotation, Lime, Starter | 1530 | 3114 | 2602 |

### 7.4.4 Crop yields

Sustainability assessment of Sanborn Field was based on the yield files dating from 1950–1989. Those yield files have been condensed into Table 7.5 to provide the reader some sense of the level of production. The study period was divided into three periods from earlier work and that division was retained.

Yields, especially on plots which received balanced fertility, tended to increase with time. This increase was a combined effect of greater genetic potential of the crop varieties and the resultant increase in fertilizer rates based upon the higher yield goals. There are insufficient plots and treatments to measure the actual magnitude of this trend. However, such a measure may not be critical because a positive trend may enhance the sustainability of a given system. The averages by time period mask a great deal of variability in annual yields within a given time period (Figures 7.1 and 7.2).

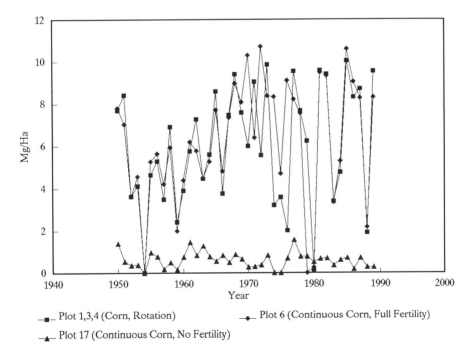

**Figure 7.1** Corn yields produced in crop rotation and in continuous culture with and without added fertility

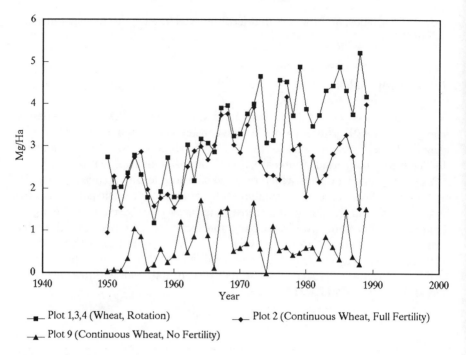

**Figure 7.2** Wheat yields produced in crop rotation and in continuous culture with and without added fertility

### 7.4.5 Weather

The Sanborn Field site is 93°20′ west longitude and 38°57′ north latitude. The continental climate is featured by a maximum monthly temperature of 31.7 °C in July with January featuring an average daily high of 2.2 °C and a low of −7.2 °C. Forty-six percent of the annual 916 mm of precipitation falls between 1 May and 30 September. The average annual precipitation during the 1950–89 period of study is given in Table 7.6.

## 7.5 PERFORMANCE OF EXPERIMENTAL PLOTS OVER TIME

A sustainable agriculture system should be economically and ecologically viable over time. In addition, the productivity of the soil resource must not depreciate over time. This particular facet of sustainability is difficult to measure using yields because changes in technology can mask soil degradation.

**Table 7.6** Annual Precipitation in Columbia, MO

| Year | Annual Precipitation (mm) | Year | Annual Precipitation (mm) |
|---|---|---|---|
| 1950 | 764 | 1970 | 1095 |
| 1951 | 1154 | 1971 | 721 |
| 1952 | 832 | 1972 | 863 |
| 1953 | 625 | 1973 | 1267 |
| 1954 | 797 | 1974 | 1100 |
| 1955 | 938 | 1975 | 1107 |
| 1956 | 663 | 1976 | 596 |
| 1957 | 811 | 1977 | 914 |
| 1958 | 933 | 1978 | 950 |
| 1959 | 774 | 1979 | 787 |
| 1960 | 704 | 1980 | 579 |
| 1961 | 1176 | 1981 | 1206 |
| 1962 | 690 | 1982 | 1189 |
| 1963 | 679 | 1983 | 1146 |
| 1964 | 839 | 1984 | 1257 |
| 1965 | 1057 | 1985 | 1377 |
| 1966 | 677 | 1986 | 1060 |
| 1967 | 837 | 1987 | 848 |
| 1968 | 925 | 1988 | 553 |
| 1969 | 1263 | 1989 | 879 |

The section will address the economic viability from two perspectives:

- output/input relationships using actual input and product prices for each of the 40 years and;
- a total factor productivity index (TFP) procedure.

The measure of economic viability will be the output/input ratio using actual prices. It reflects the short run profitability of a cropping system's net cash flow after accounting for the costs of production.

Our productivity calculations reflect market conditions during the period of this study. The output/input calculations understate the results for some years because farm subsidies tied to commodities in the form of direct payments to farmers were not considered. In short, some of the years with low factor productivity estimates may not reflect the true cash flow situation from the farm system. Units of measurement for this particular study illustrate the inputs/outputs associated with one acre.

A second technique of sustainability calculation pertains to the use of a TFP index. The Tornquist index was selected because of its superiority to the arithmetic-based indexes. (See Chapter 3.) The year 1950 was selected as the base year, and subsequent indexes are considered as deviations from the base year permitting detection of trends regarding sustainability.

Calculations were based on the following procedures and assumptions.

- Actual prices for plant nutrients among years were used.
- Prices received for produce were averages for the state of Missouri.
- Production technology (and associated costs) used by typical central Missouri crop farmers were used in productivity calculations. Labour and machinery operating costs per acre were either provided or calculated from data provided from farm record participants' summaries. Missouri farmers participate in a 'mail-in' record program which permits individual financial evaluations and compilation into statewide records on an annual bases. Early years of farm records were variable with regard to specifics, but data permitted labour and machinery costs calculations on an acre basis. (Missouri Farm Business Summaries, Missouri Cooperative Extension Service).
- Cash rental rates were used as proxy for land costs (US Department of Agriculture).
- Yields on Sanborn Field were assumed comparable to those of farmers using typical technologies of farming in central Missouri.

Figures 7.3–7.8 show the annual output/input ratio by year. The overall output/input for the fully treated corn–wheat–red clover rotation (Plots 1, 3, and 4) shows that during the early years of the period (1954–64), costs of production exceeded value of output (Figure 7.3). The reverse holds true for about the next 15 years when value of output exceeds the costs, i.e., the output/input ratio exceeds the numerical value of 1.

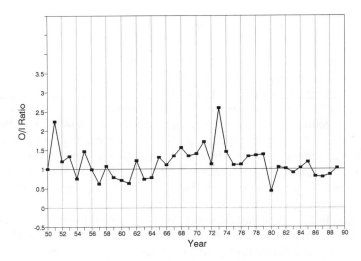

**Figure 7.3** Average output/input ratios of a corn–wheat–clover rotation with each crop grown every year under full fertility (plots 1, 3 and 4).

## Performance of experimental plots over time 123

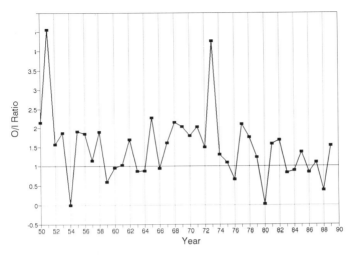

**Figure 7.4** Output/input ratios of corn grown in rotation with wheat and clover under full fertility (plots 1, 3 and 4)

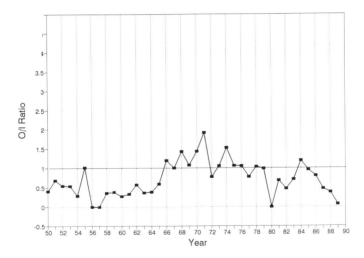

**Figure 7.5** Output/input ratios of red clover grown in rotation with corn and wheat under full fertility (plots 1, 3 and 4)

A breakdown of the data from the corn–wheat–clover plots into the individual crops was accomplished to better identify the components of the rotation output/input ratio (Figures 7.4, 7.5 and 7.6). The output/input ratio for corn was more variable year-to-year than for wheat, because the weather during the winter wheat growing period is less subject to wide swings in precipitation and temperature than in the summer corn growing season.

Clover had more years when the output/input ratio was below 1 than either corn or wheat. This is probably because of poor economic and weather conditions. Early in the study period, hay was priced locally and had a low commercial price but recent developments have made hay marketing a more profitable enterprise. Also, wheat serves as a nurse crop for clover establishment and vigorous wheat tends to shade clover seedlings, reducing the stand. In addition, the clover production period coincides with that of corn so the same weather stresses that affect corn also affect clover.

Corn tended to dominate the profitability calculations as reflected by the output/input ratio for given crop rotations. Corn is a major cash crop in midwestern US agriculture and it merits further sustainability evaluation. The output/input ratios of corn from the corn–wheat–clover rotation, continuous corn and the corn–oat–wheat–clover rotation are compared in Figures 7.7 and 7.8. All output/input ratios tended to be > 1 under full fertility. The output/input ratio for the four-year rotation had less fluctuation but this rotation was in corn only one in four years. Corn was, of course, grown every year continuously and also every year on one plot of the three-year rotation.

The period 1960–77 tended to have favourable weather for corn production. There was no corn harvested in 1954 and 1980 because of crop failure due to adverse weather. Thus, an outpur/input ratio in one individual year suggests growing no crop would have been better than trying to grown corn. However, over the long term, corn which received adequate nutrition had an output/input ratio over 1 and was sustainable (Figure 7.7).

Emphasis was placed upon corn (maize) rather than wheat because the output/input ratio measures suggested corn was sustainable whether grown continuous-

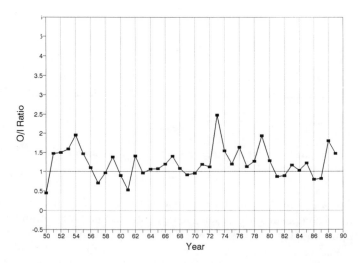

**Figure 7.6** Output/input ratios of winter wheat grown in rotation with clover and corn under full fertility (plots 1, 3 and 4)

ly or in rotation with other crops. Wheat, on the other hand, did not fare well in continuous cropping due to recurring disease problems. In addition, yields were highly related to added plant nutrients (see Table 7.1).

As pointed out earlier, continuous corn may be sustainable if provided with balanced nutrition. Corn with no plant nutrient applications consistently failed to

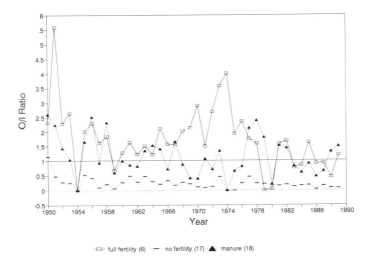

**Figure 7.7** Output/input ratios of continuous corn grown with different fertility levels (6, 17 and 18)

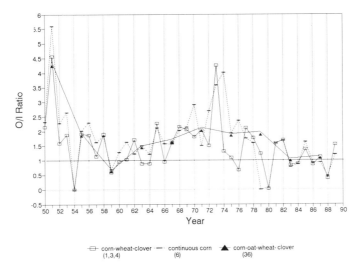

**Figure 7.8** Output/input ratios of continuous corn and corn grown in rotation with full fertility (plots 1, 3, 4, 6, and 36)

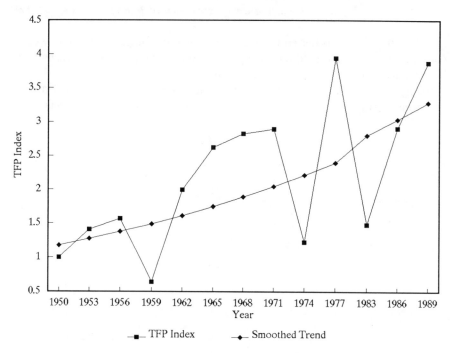

**Figure 7.9** TFP index of corn in rotation with wheat and clover under full fertility (plot 1)

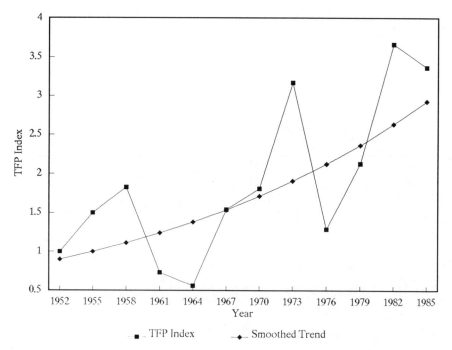

**Figure 7.10** TFP index of corn in rotation with wheat and clover under full fertility (plot 3).

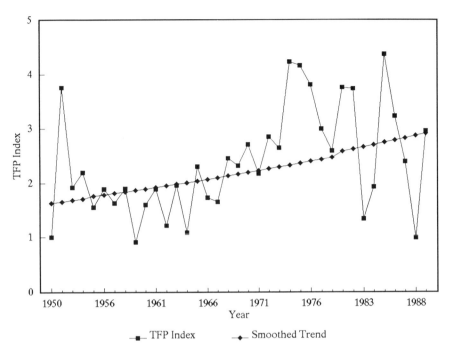

**Figure 7.11** TFP index of continuous corn grown with full fertility (plot 6)

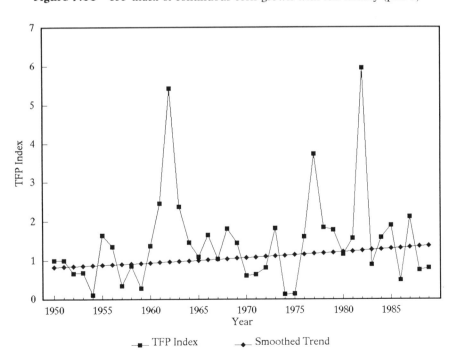

**Figure 7.12** TFP index of continuous corn grown with no added fertilizer (plot 17)

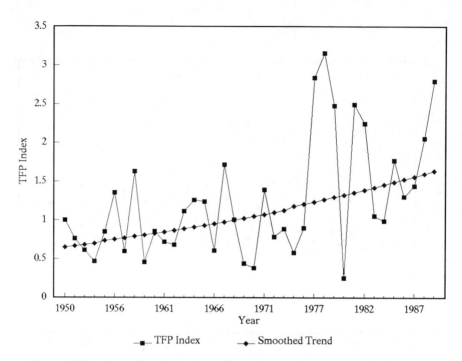

**Figure 7.13** TFP index of continous corn grown with added manure (plot 18)

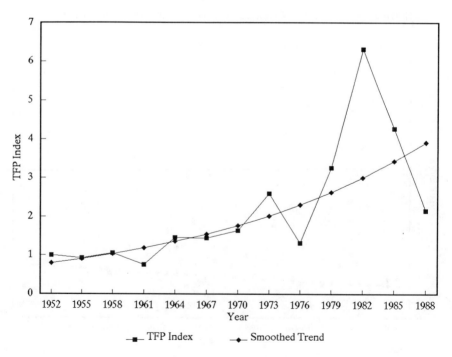

**Figure 7.14** TFP index of corn grown in rotation with wheat and clover under full fertility (plot 26)

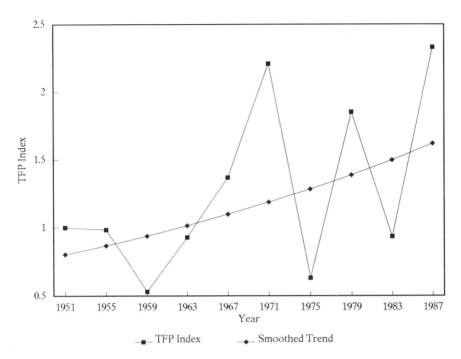

**Figure 7.15** TFP index of corn grown in rotation with wheat, oats and clover with manure application (plot 34)

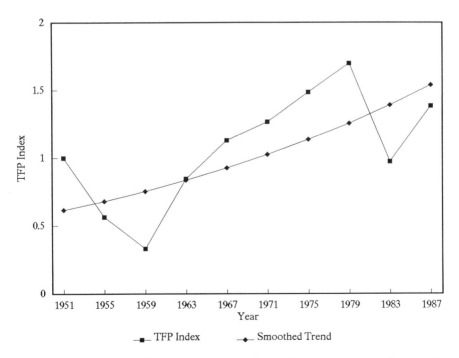

**Figure 7.16** TFP index of corn grown in rotation with wheat, oats and clover under full fertility

have benefits in excess of costs. (Figure 7.7) Corn with manure applications compared favourably with corn grown with full fertilizer treatment until the mid-1960s, at which time full fertilizer treatment yields were substantially higher. This resulted from increased nitrogen use on the fully treated plots and higher yield potentials of the corn hybrid. During the last decade of the experiment, the output/input ratios of the two treatments tracked in a comparable fashion.

Essentially, every year that the plots were in clover (starting in 1951) the output/input ratio fell below 1. This result from clover was discussed above. What cannot be measued from the data is the benefit clover had upon the following corn crop.

To gain further insight into the benefits associated with corn grown in rotation, in contrast to continuous corn, output/input ratios were compared under full fertility nutrient programmes. Figure 7.8 shows these comparisons. Any benefit from corn grown in rotation appears limited. However, a few of the more recent years had higher output/input ratios when corn was grown in rotation. We conclude that a balanced fertility programme contributes to sustainability of corn production either in rotation with other crops or in continuous culure.

The finding that rotations had a relatively large number of years when output/input ratios were less than 1 (costs exceeds returns) must be interpreted with caution because the feed value of red clover has not been included. Only the cropping patterns in the early years of the experiments were systems in which animals were a major component. For example, red clover was feed for ruminants and pasture for swine, and it played a major role in farms where clover-fixed N preceded corn planting. As the use of fertilizer nitrogen increased over the four decades the contribution from nitrogen fixed by the sparse clover had increasingly less impact.

The contribution of red clover to other enterprises was not accounted for in our calculations—only the value of red clover hay that was actually sold. In addition, hay markets, until the very recent years, were rudimentary in nature and were often limited to transactions among neighbours.

## 7.6 AN INDEX APPROACH

Technology advances generally displace other resources (especially labour) in production activities. Quantifying the impact of technical change and how it affects production growth is of greater importance. An alternative to measuring the traditional real value of added output and average labour productivities is that of measuring total factor productivity growth. Use of the Tornquist Index (Gollop and Jorgenson, 1980), to estimate the total factor productivity growth has become more widespread. This approach enables one to observe the rate of technical change without estimating the production function or any indirect objective function. (See Chapter 3.) The year of 1950 was arbitrarily selected as the base year; hence, subsequent TFP indexes are interpreted as deviations from the base year value of 1.

This part of the paper will focus only on corn with respect to sustainability as reflected by index calculations. Years with extremely low yields were considered as statistical outliers and not used in the index calculations. The TFP indexes still show considerable variation because of yield extremes. This poses a problem of ascertaining whether there has been a trend for total factor productivity changes associated with corn production. In order to make a judgement on this matter, TFP Indexes were regressed on time to provide coefficients that might calculate a 'smoothed' trend to assess sustainability over time.

The upward sloping trends for the various plots suggest that technology advances have been sucessful in increasing total factor productivity (Figures 7.9–7.16). Limited observations of corn grown in rotation suggest that total factor productivity growth exceeds growth rates obtained with continuous corn. In addition, the productivity rate of change is diminished as fertility levels deviate from the optimum.

## 7.7 CONCLUSIONS

The use of both the output/input ratio and the Tornquist index (TFP) demonstrated that balanced nutrient management influences the sustainability of crop production. Continuous culture of wheat without the use of fungicides and insecticides was not sustainable even when adequate nutrition was supplied.

Corn production was sustainable in continuous culture. Sustainability of corn grown in rotation was influenced by the variability of the red clover hay output as well as weather induced wide swings in corn yields.

The decade from 1964–73 gave the most production due to favourable weather for all crops. Both adverse economic conditions and widely fluctuating yields from 1974–89 made the sustainability indices less positive. The TFP index calculations showed that total factor productivity has increased over the last 40 years but may have stabilized during the last decade.

Crop rotation during the four-decade study period maintained the productivity of the soil. This was indicated by both yields and soil organic matter when balanced nutrition was used.

Midwestern agriculture, particularly in the early years of this study, was characterized by farming systems that had animals playing a major role in the total farm business. Therefore, any inferences about the welfare of farm families from only the study of the cropping systems must be interpreted with caution. The wide swings in weather conditions greatly influenced yields and thus a major limitation of this study was that these weather conditions were not considered in a multivariate context when examining total factor productivity changes over time. An econometric approach that considers climatic conditions as an explanatory variable of yields should give further insight into sustainability issues. In addition, soil property changes over time were not considered in total factor productivity calculations. Sustainability measures that will account for soil property changes warrant attention in future research.

# 8
# Major Cropping Systems in India

K. K. M. Nambiar

Indian Agricultural Research Institute, New Delhi, India

## 8.1 INTRODUCTION

The world has witnessed unprecedented advances in the development and application of 'science-based agriculture production practices' since the 1950s. Global food production has increased 2.6-fold, while the production in India alone has increased 3.5-fold since the mid 1960s. The worldwide concern is whether such increases are sustainable. This does not seem to be misplaced as global food production has shown a declining trend since the beginning of the 1980s. Moreover, it is estimated that the world population will be around 6.4 billion by the year 2000 AD. The population of India alone might be around 1.0 billion. Thus, world agriculture would be called upon to produce as much food as has been produced in the entire 1200-year history to meet the food demand of the next two to four generations (Freeman, 1990).

The term 'sustainability' has been defined in many different ways. According to FAO (1988), 'Sustainable development is the management and conservation of the natural resource base and orientation of technological institutional change in such a manner as to ensure the attainment and continued satisfaction of human needs for present and future. Such sustainable development (in agriculture, forestry and fishery sectors) conserves land, water, plant and animal genetic resources, is environmentally non-degrading, technically appropriate, economically viable and socially acceptable'. The American Society of Agronomy defined sustainable agriculture as 'One that, over the long-term enhances environmental quality and the resource base on which the agriculture depends,

---

*Agricultural Sustainability: Economic, Environmental and Statistical Considerations.*
Edited by V. Barnett, R. Payne and R. Steiner © 1995 John Wiley & Sons Ltd

provides for human fibre and food needs, is economically viable, and enhances the quality of life for farmers and society as a whole'. In crop production, sustainability is regarded as maintaining a given production level, over time, and enhancing production to accommodate an expanding population (Anders, 1990). Thus, soil degradation, loss of genetic diversity, climatic changes, pests and diseases, nutrient stresses, etc. lead to loss of productivity and adversely affect the sustainability of the production systems. Maintaining sustained productivity in the years to come is a colossal task both in magnitude and complexity. To assess the long-term sustainability of modern farming, the Indian Council of Agricultural Research sponsored the All-India Coordinated Research Project on Long Term Fertilizer Experiments during the IV Plan period (September 1970) at 11 selected centres, representing the major soil climatic regimes of India. The effects of plant nutrients on the long-term sustainability of four major cropping systems on alluvial soil (Ludhiana, Punjab), medium black soil (Jabalpur, Madhya Pradesh), laterite soil (Bhubaneswar, Orissa) and foothill soil (Pantnagar, Uttar Pradesh) are described in this chapter.

## 8.2 CLIMATE

The climate of Ludhiana (Punjab) and Jabalpur (Madhya Pradesh) is semi-arid, while it is humid and subhumid tropical in Bhubaneswar (Orissa) and Pantnagar (Uttar Pradesh). The mean annual precipitation varies from 800 to 900 mm and 1000 to 1200 mm, respectively, at Ludhiana and Jabalpur. It is 1400 mm and 1449 mm, respectively, at Pantnagar and Bhubaneswar. The mean annual minimum and maximum air temperature from January to June varies from 4.3–43.8 °C and 5.9–42.8 °C, respectively, at Ludhiana and Jabalpur. It ranges from 14.0 °C in December to 37.5 °C in June at Bhubaneswar and 3.4 °C in January to 39.2 °C in May at Pantnagar.

## 8.3 SOILS

Soils of the experimental sites at Ludhiana (Punjab), Jabalpur (Madhya Pradesh), Bhubaneswar (Orissa) and Pantnagar (Uttar Pradesh) have been classified as alluvial, medium black, laterite and *tarai* (foothill) soils, respectively.

The soil textures of the plough layer (0–22.5 cm) at the experimental sites of Ludhiana and Bhubaneswar are loamy sand. However, the soil texture at Bhubaneswar below 22.5 cm depth of the profile becomes clay loam. Similarly, the soil textures are clay and silty clay loam, respectively, at Jabalpur and Pantnagar. The initial basic characteristics of the soils are given in Table 8.1.

**Table 8.1** Some basic characteristics of the soils in the long-term fertilizer experiments

| Location | Soil group | Taxonomic Class | Texture | Dominant clay minerals | pH | CEC (cmol(p$^+$)/kg) | Org.C % | *Available nutrients (mg/kg) N$^a$ | P$^b$ | K$^c$ |
|---|---|---|---|---|---|---|---|---|---|---|
| Ludhiana | Alluvial | Ustochrept | Loamy sand | Illite | 8.2 | 5.1 | 0.21 | 48.5 | 4.5 | 44.0 |
| Jabalpur | Medium black | Chromustert | Clay | Montmorillonite | 7.8 | 49.0 | 0.57 | 113.0 | 4.0 | 185.0 |
| Bhubaneswar | Laterite | Haplaquept | Loamy sand | Kaolinite | 5.5 | 4.0 | 0.27 | — | 15.5 | 12.5 |
| Pantnagar | Foothill | Hapludoll | Silty clay loam | Chlorite, illite | 7.3 | 20.0 | 1.48 | 254.5 | 9.0 | 67.5 |

*Elemental form:
[a] Alkaline permanganate extractable-N.
[b] Olsen-P (Bicarbonate extractable-P).
[c] N NH$_4$OAc-extractable K.

## 8.4 EXPERIMENTAL DESIGN

The experimental design consists of 10 or 11 treatments laid out in randomised blocks with 4 replications (3 replications at Bhubaneswar). Each centre follows a typical multiple cropping system in which high yielding varieties are raised under irrigated conditions. Fertilizer doses were decided on initial soil-test values and local recommendations (Table 8.2). Farmyard manure (FYM) is applied at 10–15 t/ha annually to the *kharif* (monsoon) crop and zinc sulphate (as per the need) to the *rabi* (winter) crop in the respective treatments. The plot size ranges from 160 to 300 m$^2$. Full particulars of the experiments have been reported by Nambiar and Ghosh (1984).

Plotwise samples (0–15 cm) have been tested periodically to monitor changes in the available nutrient status. Plant samples collected at harvest were analysed to determine the nutrient uptake. Five treatments, namely 100% optimal NPK (Sulphur-free sources) (NPK), 100% optimal NPK (NPKS), 100% NPK + Zinc (NPKSZn), 100% NPK + FYM (NPKSM) and control (unmanured) were selected out of 10 or 11 treatments for evaluating the biological sustainability of four major cropping systems (Table 8.2) under balanced and optimal use of major nutrients (NPK), with and without secondary and micronutrient elements or

**Table 8.2** Cropping system and optimum NPK dose (100%) in long-term fertilizer experiments at the selected locations

| Location | Soil group | Annual crop rotation and nutrients added based on 100% optimal NPK dose (kg N-P$_2$O$_5$-K$_2$O/ha) | | |
|---|---|---|---|---|
| | | 1st crop | 2nd crop | 3rd crop |
| Ludhiana (Punjab) | Alluvial | Maize (150-75-75) | Wheat (150-75-37.5) | Cowpea fodder (20-40-20) |
| Jabalpur (Madhya Pradesh) | Medium black | Soybean (20-80-20) | Wheat (120-80-40) | Maize fodder (80-60-20) |
| Bhubaneswar (Orissa) | Laterite | Rice (100-60-60) | Rice (100-60-60) | — |
| Pantnagar (Uttar Pradesh) | Foothill | Rice (120-60-45) | Wheat (120-60-40) | Cowpea fodder (0-0-0) |

*Treatment details*
1. 50% of optimal NPK (based on initial soil tests and local recommendations) (50% NPKS)
2. 100% of optimal NPK (based on initial soil tests and local recommendations) (NPKS)
3. 150% of optimal NPK (based on initial soil tests and local recommendations) (150% NPKS)
4. 100% of NPK and handweeding (instead of chemical control in all other treatments) (NPKSHW)
5. 100% of NPK + Zinc/lime (NPKSZn)/(NPKSL)
6. 100% of optimal NP (NP)
7. 100% of optimal N (N)
8. 100% of NPK + FYM (10–15 t/ha) (NPKSM)
9. 100% of NPK (Sulphur-free sources) (NPK)
10. Control (no fertilizer and manure) (unmanured) + Additional treatment, if any of local interest.

**Table 8.3** Yearwise standard error of difference of two treatment means for the grain yield of crops under ICAR Coordinated Long Term Fertilizer Experiments (t/ha)

| Year | Alluvial (Ludhiana) | | Medium black (Jabalpur) | | Laterite (Bhubaneswar) | | Foothill (Pantnagar) | |
|---|---|---|---|---|---|---|---|---|
| | Maize | Wheat | Soybean | Wheat | Rice[†] | Rice[‡] | Rice | Wheat |
| 1971–72 | — | 0.244 | — | — | — | 0.270 | — | — |
| 1972–73 | 0.200 | 0.161 | 0.255 | 0.250 | 0.402 | 0.491 | 0.123 | 0.151 |
| 1973–74 | 0.141 | 0.146 | 0.175 | 0.213 | 0.345 | 0.427 | 0.229 | 0.195 |
| 1974–75 | 0.122 | 0.073 | — | 0.115 | 0.248 | 0.375 | 0.225 | 0.267 |
| 1975–76 | 0.097 | 0.073 | 0.099 | 0.448 | 0.317 | 0.318 | 0.451 | 0.218 |
| 1976–77 | 0.088 | 0.083 | 0.211 | 0.190 | 0.506 | 0.396 | 0.537 | 0.365 |
| 1977–78 | 0.073 | 0.063 | 0.216 | 0.102 | 0.214 | 0.389 | 0.448 | 0.426 |
| 1978–79 | 0.058 | 0.093 | 0.216 | 0.102 | 0.327 | 0.399 | 0.371 | 0.376 |
| 1979–80 | 0.068 | 0.122 | 0.192 | — | 0.375 | 0.276 | 0.768 | 0.192 |
| 1980–81 | 0.063 | 0.113 | 0.127 | 0.136 | 0.330 | 0.146 | 0.724 | 0.267 |
| 1981–82 | 0.084 | 0.069 | 0.102 | 0.117 | 0.266 | 0.202 | 0.396 | 0.182 |
| 1982–83 | 0.088 | 0.088 | 0.112 | 0.239 | 0.183 | 0.201 | 0.625 | 0.305 |
| 1983–84 | 0.092 | 0.099 | 0.110 | 0.180 | 0.131 | 0.117 | 0.313 | 0.173 |
| 1984–85 | 0.080 | 0.113 | 0.096 | 0.122 | 0.211 | 0.257 | 0.382 | 0.584 |
| 1985–86 | 0.090 | 0.081 | 0.103 | 0.119 | 0.255 | 0.199 | 0.452 | 0.414 |
| 1986–87 | 0.080 | 0.104 | — | 0.167 | 0.264 | 0.276 | 0.357 | 0.360 |
| 1987–88 | 0.064 | 0.081 | 0.137 | 0.199 | 0.244 | 0.242 | 0.184 | 0.118 |
| 1988–89 | 0.097 | 0.076 | 0.211 | 0.234 | 0.316 | 0.241 | 0.167 | 0.320 |
| 1989–90 | 0.072 | 0.115 | 0.140 | — | 0.365 | 0.168 | 0.209 | 0.186 |
| 1990–91 | — | — | 0.131 | 0.184 | 0.230 | 0.181 | 0.205 | 0.183 |

[†] Monsoon rice.
[‡] Winter rice.

farmyard manure (FYM). The mean annual grain yields as well as three-year moving average grain yields were plotted over the years (1971–91) for the unmanured, NPK, NPKS, NPKSZn and NPKSM treatments. Similarly, the mean values were used for plotting the changes in physicochemical properties of the soils. The yield data as well as the soil properties were subjected to statistical analysis and have been presented in the respective annual reports of the project since 1972. The yearwise standard error of difference of two treatment means in respect of grain yield of crops is given in Table 8.3. The pooled analysis of the yield data over the years (1971–89) was reported in the annual reports 1987–88 and 1988–89 (Nambiar, 1992b).

## 8.5 SUSTAINABILITY ON ALLUVIAL SOIL AT LUDHIANA, PUNJAB

Trends in the yield of maize and wheat crops, organic carbon and available soil N, P, K and Zn as affected by long-term use of plant nutrients on the light alluvial sandy soil of Ludhiana (Punjab) are depicted in Figures 8.1–8.9.

### 138   *Major cropping systems in india*

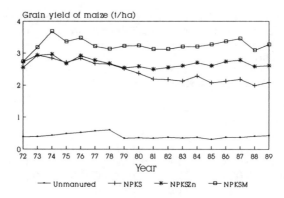

**Figure 8.1**   Yield trend of maize on alluvial soil in ICAR LTF experiments

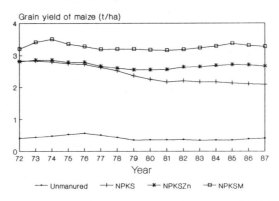

**Figure 8.2**   Yield trend (three-year moving average) on alluvial soil in ICAR LTF experiments

### 8.5.1   Yield trend

Yield trend of maize showed that its productivity was more or less maintained with the recommended doses of mineral NPKS fertilizers in the initial nine-year period. Subsequently, there was a declining trend (Figures 8.1 and 8.2) which was arrested with the application of Zn (once in 3 to 8 years, see Section 8.5.7). Incorporation of farmyard manure (M) along with S-carrying mineral NPK fertilizers enhanced the productivity of maize over the NPKSZn treatment (Nambiar and Abrol, 1989; Nambiar, 1989; Nambiar, 1990; Nambiar, 1991a, b).

The productivity of wheat was maintained throughout the experimental period (1971–89) with the recommended doses of mineral NPK fertilizers (Figures 8.3 & 8.4). The residual effect of farmyard manure (M) was found to be negligible on wheat productivity. Similarly, no effect of Zn was found.

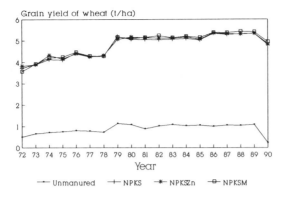

**Figure 8.3** Yield trend of wheat on alluvial soil in ICAR LTF experiments

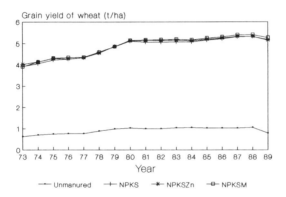

**Figure 8.4** Yield trend (three-year moving average) on alluvial soil in ICAR LTF experiments

### 8.5.2 Soil reaction

Soil pH has a manifold influence on all the chemical, biological and physical processes in soils and on nutrient availability to crops. Interestingly, no significant effect on soil pH was noticed as a result of continuous manuring and cropping of maize–wheat–cowpea fodder rotation over the years 1971–89 (Nambiar, 1992b).

### 8.5.3 Soil organic carbon

The importance of soil organic matter is well recognised in maintaining soil fertility and productivity and its maintenance in soils is of the utmost concern in

## 140  Major cropping systems in india

intensive cropping systems. It is of great interest to note that a small improvement in soil organic matter was noticed even on the unmanured plots. A similar improvement was also observed with recommended doses of mineral NPKS fertilizers (Nambiar, 1985b). An increasing trend was obtained, however, after nine annual cropping cycles, with the integrated use of farmyard manure and mineral NPK fertilizers (NPKSM treatment); see Figure 8.5.

### 8.5.4  Available soil nitrogen

A small improvement in available soil N was noticed even for the unmanured plots. A similar improvement in available soil N was also observed with recom-

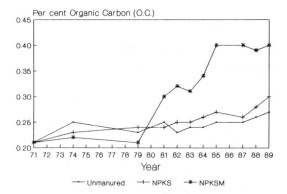

**Figure 8.5** Trend of organic carbon in maize–wheat–cowpea (fodder) rotation on alluvial soil in ICAR LTF experiments

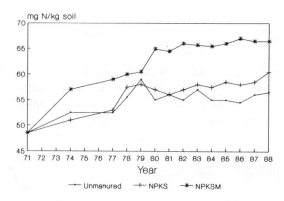

**Figure 8.6** Trend of average N in maize–wheat–cowpea (fodder) rotation on alluvial soil in ICAR LTF experiments

mended doses of S-carrying mineral NPK fertilizers but with a slightly higher rate of increase. The available soil N steadily improved with the integrated use of organic manure and mineral NPK fertilizers (NPKSM treatment) but levelled off after 10 annual cropping cycles (Figure 8.6) indicating an equilibrium between application and uptake of N by crops (Nambiar, 1985b).

### 8.5.5 Available soil phosphorus

The available soil P (Olsen's P) showed a steady improvement with recommended doses of mineral NPKS fertilizers. A similar trend was also noticed with the integrated use of organic manure and mineral NPK fertilizers (NPKSM treatment). The available soil P declined sharply in the absence of P application in the initial three-year period but subsequently evened out (Figure 8.7) indicating a constant release of P from the native source at which the reduction in P was found to be extremely slow (Nambiar 1992b).

### 8.5.6 Available soil potassium

A steady rise in available soil K was noticed in the first three year period for the NPKSM and NPKS treatments but this flattened off subsequently indicating an equilibrium state between K uptake by crops and fertilizer K applied, including K release from the soil (Figure 8.8). However, a downward trend in available soil K was observed for the NPKS treatment after 14 annual cropping cycles (Figure 8.8) implying an imbalance between K uptake by crops and fertilizer

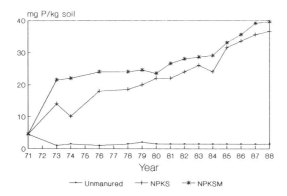

**Figure 8.7** Trend of average P in maize–wheat–cowpea (fodder) rotation on alluvial soil in ICAR LTF experiments

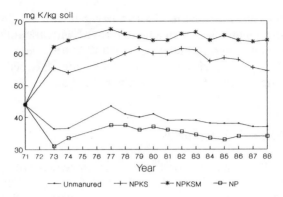

**Figure 8.8** Trend of average K in maize–wheat–cowpea (fodder) rotation on alluvial soil in ICAR LTF experiments

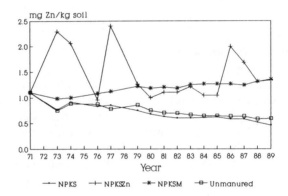

**Figure 8.9** Trend of average Zn in maize–wheat–cowpea (fodder) rotation on alluvial soil in ICAR LTF experiments

K applied, including K release from the soils. A drastic reduction in available soil K was noticed in the first three-year period in the absence of K application. Subsequently, it more or less attained an equilibrium state without any sharp-change in available soil K (Nambiar, 1985a).

### 8.5.7 Available soil zinc

Available soil Zn (DTPA-extractable Zn) showed a steady declining trend (Figure 8.9) in the absence of Zn or farmyard manure application over the years 1971–89. However, it maintained the initial Zn level with the incorporation of farmyard manure (Nambiar and Abrol, 1989). The level of available soil Zn increased with application of Zn (50 kg/ha of $ZnSO_4 \cdot 7H_2O$) once in 3–8 years as the available soil Zn decreased to around 1.0 mg/kg.

### 8.5.8 Discussion of the results at Ludhiana

From the yield trend of maize (Figures 8.1 & 8.2), it can be seen that the productivity of maize with balanced application of S-carrying mineral N, P and K fertilizers showed a declining trend on light alluvial sandy (Ludhiana) after nine annual cropping cycles. This was found to be associated with the emerging Zn deficiency (Nambiar and Abrol, 1989; Nambiar, 1991a); the drawdown in available soil Zn was from the initial level of 1.10 to around 0.68 mg/kg. However, significant reduction in maize productivity was noted when the available soil Zn decreased to around 0.60 mg/kg. Application of zinc (50 kg/ha of $SO_4 \cdot 7H_2O$) once in 3–8 years (along with mineral NPKS fertilizers) restored the initial productivity of maize. It would therefore appear that plants need not only adequate quantities of NPK nutrients but also secondary and micronutrient elements, especially S and Zn which are often found to be in deficit on light alluvial sandy soils.

Similarly, incorporation of farmyard manure (M) along with mineral NPKS fertilizers not only maintained the initial productivity of maize but also improved over the NPKSZn treatment (Figures 8.1 and 8.2.) This could be ascribed to its providing an adequate quantity of available Zn thereby correcting the deficiency on the one hand and enhancing the efficiency of use of the NPK fertilizers on the other. This was also seen in the increasing trends of organic carbon, available soil N, P and K over the years 1971–89, with integrated use of organic manure and mineral NPKS fertilizers (NPKSM treatment).

The productivity of wheat was maintained throughout the experimental period (1971–89) with recommended doses of S-carrying mineral NPK fertilizers alone (Figures 8.3 and 8.4). It is widely known that maize is more susceptible to Zn deficiency than wheat. Moreover, wheat has a deeper root system than maize, which makes it able to exploit greater soil volume, thus deriving an adequate quantity of Zn for meeting its Zn demand (Nambiar, 1990).

The declining trend in maize productivity was therefore found to be associated with the deficiency of Zn which could be corrected through periodical application of Zn (as per the need). Integrating organic manure (15 t/ha/yr) with mineral NPK fertilizers was also found to be quite promising in arresting the deterioration in productivity of maize, through the correction of marginal deficiencies of certain secondary and micronutrient elements in the course of mineralization of organic manures on the one hand, and its beneficial influence on the physical and biological properties of the soil on the other, thus leading to improved soil fertility and productivity.

## 8.6 SUSTAINABILITY ON MEDIUM BLACK SOIL AT JABALPUR, MADHYA PRADESH

The trends in soy bean and wheat yield, organic carbon, available soil N, P, K and S as influenced by long-term use of plant nutrients on medium black soil at Jabalpur (Madhya Pradesh) are shown in Figures 8.10–8.18.

### 8.6.1 Yield trend

Yield trend of soy bean indicated a deterioration in productivity with the continuous use of S-free mineral NPK fertilizers after two annual cropping cycles. Reduction in soy bean productivity became significant after 12 annual rotations

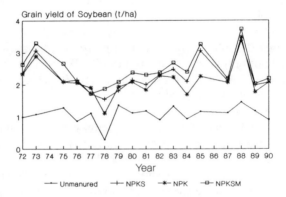

**Figure 8.10** Yield trend of soy bean on medium black soil in ICAR LTF experiments

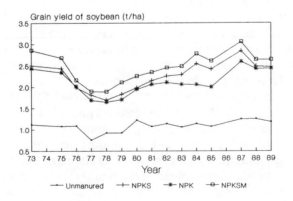

**Figure 8.11** Yield trend (three-year moving average) on medium black soil in ICAR LTF experiments

(Figures 8.10 and 8.11). Marked improvement in yield of soy bean was obtained with the use of S-containing NPK fertilizers after 12 annual rotations (Nambiar and Abrol, 1989). A similar improvement in yield was also noted with the integrated use of farmyard manure and mineral NPKS fertilizers (NPKSM treatment). Grain yield of soy bean was more less stable around 1.0 t/ha on unmanured plots in the initial 16-year period but thereafter exhibited a diminishing trend with a reduction in soil fertility in the absence of the addition of plant nutrients.

Wheat yield showed a downward trend in the initial five-year period with recommended doses of S-free mineral NPK fertilizers which stabilised in the subsequent four-year period. An upward trend in wheat productivity was noticed with S-free mineral NPK fertilizers after nine annual rotations (Figures 8.12 and 8.13). The improvement in wheat productivity was found to be related to the

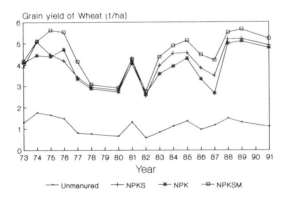

**Figure 8.12** Yield trend of wheat on medium black soil in ICAR LTF experiments

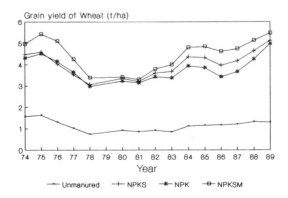

**Figure 8.13** Yield trend (three-year moving average) on medium black soil in ICAR LTF experiments

improvement in P fertility after 10 annual rotations (Figure 8.16). Wheat productivity with the continuous use of S-containing NPK fertilizers improved remarkably over S-free NPK fertilizers after 11 annual rotations indicating incipient S deficiency (Nambiar, 1988). The highest wheat productivity was maintained with the integrated use of farmyard manure and mineral NPKS fertilizers (NPKSM treatment) which could be ascribed to correction of marginal deficiencies of secondary and micronutrient elements in the course of mineralization of organic manure.

### 8.6.2 Soil reaction

There was no significant change in soil pH with continuous use of mineral NPKS fertilizers over the years 1981–89. Similarly, there was no effect on soil pH with integrated use of organic manure and mineral NPK fertilizers (Nambiar, 1992b).

### 8.6.3 Soil organic carbon

A small improvement in soil organic carbon was noticed with recommended doses of S-containing mineral NPK fertilizers. Soil organic carbon showed an increasing trend with the integrated use of farmyard manure and mineral NPKS fertilizers (NPKSM treatment) in the initial four-year period and levelled off in the subsequent nine-year period. However, it showed a steady improvement after 15 annual rotations (Figure 8.14). Little change in soil organic carbon was noticed in the case of the unmanured treatment. This appeared to be due to a smaller addition of crop residues and root biomass as a result of poor crop growth in the absence of adequate application of plant nutrients (Nambiar, 1992b).

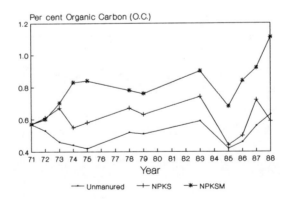

**Figure 8.14** Trend of organic carbon in soy bean–wheat–maize (fodder) rotation on medium black soil in ICAR LTF experiments

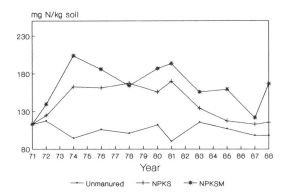

**Figure 8.15** Trend of average N in soy bean–wheat–maize (fodder) rotation on medium black soil in ICAR LTF experiments

### 8.6.4 Available soil nitrogen

The available soil N with recommended doses of mineral NPKS fertilizers steadily improved in the initial four-year period, attained an equilibrium in the subsequent seven-year period and thereafter showed a diminishing trend (Nambiar, 1992b). A similar trend was also observed for the integrated use of farmyard manure and mineral NPKS fertilizers (NPKSM treatment). With the NPKS treatment, available soil N indicated an upward trend after 15 annual rotations. In contrast, for the unmanured treatment the trend in available soil N showed a sharp decline in the initial four-year period and then attained approximate equilibrium (see Figure 8.15).

### 8.6.5 Available soil phosphorus

The available soil P did not show any appreciable change with recommended doses of mineral NPKS fertilizers in the initial 10-year period but subsequently indicated a steadily increasing trend (Figure 8.16). A similar trend was also observed with the integrated use of organic manure and mineral NPKS fertilizers (NPKSM treatment) but the rate of improvement was more marked with the NPKSM treatment than with the NPKS treatment. Lack of change in available P with NPKS and NPKSM treatments in the initial 10-year period seems to be associated with the P fixation in the soil (Nambiar and Abrol, 1988).

Medium black soil was found to be highly deficient in available P (Olsen's P) and contained less than 5 mg P/kg. It would therefore appear that the available soil P showed an increasing trend after satisfying the P fixation capacity of the soil. There was no change in available soil P with the unmanured treatment throughout the experimental period (1971–89), indicating an equilibrium state in the absence of P addition.

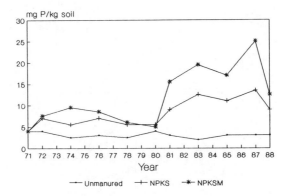

**Figure 8.16** Trend of average P in soy bean–wheat–maize (fodder) rotation on medium black soil in ICAR LTF experiments

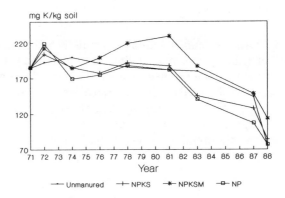

**Figure 8.17** Trend of average K in soy bean–wheat–maize (fodder) rotation on medium black soil in ICAR LTF experiments

### 8.6.6 Available soil potassium

The trend in available soil K exhibited more or less an equilibrium state with recommended doses of S-containing mineral NPK fertilizers in the initial 11-year period but thereafter declined steadily (Nambiar, 1985a; Nambiar and Ghosh, 1985; Nambiar and Ghosh, 1986). However, the available soil K showed a steady improvement with integrated use of farmyard manure and mineral NPKS fertilizers (NPKSM treatment) in the initial 11-year period but again declined in the subsequent period. The trend in available soil K with the NP and NPKS treatments was more or less similar on medium black soil (Jabalpur) due to its high native K status (Figure 8.17). Nevertheless, the rate of reduction in available soil K was greater for the NP treatment than for the NPKS treatment.

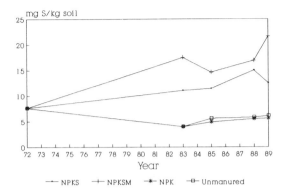

**Figure 8.18** Trend of avergae S in soy bean–wheat–maize (fodder) rotation on medium black soil in ICAR LTF experiments

### 8.6.7 Available soil sulphur

The trend in available S indicated a gradual declining trend with S-free NPK fertilizers. However, it steadily improved with the application of S-containing NPK fertilizers (Nambiar, 1988). A greater improvement was obtained with the incorporation of farmyard manure along with S-containing NPK fertilizers (Figure 8.18). Deficiency of S became evident when the available S (Morgan's-extractable S) dropped from 7.8 mg/kg to around 5 mg/kg.

### 8.6.8 Discussion of the results at Jabalpur

Yield trend of soy bean indicated an overall reduction in the productivity after two annual rotations on medium black soil (Jabalpur) irrespective of treatment (Figure 8.10). However, the highest reduction in soy bean yield was noticed in the seventh year of experimentation, that is during 1978–79, followed by an upward trend in yield in the subsequent period. The upward trend in the yield after seven annual rotations could be ascribed to improvement in P fertility with continuous application of P fertilizers (Nambiar and Abrol, 1988). The medium black soil was deficient in available soil P (less than 5 mg/kg) which improved after 10 annual rotations, satisfying the P fixation capacity of the soil and leading to improved soil fertility and productivity.

The lowest soy bean yield recorded during 1978–79 was associated with drought conditions. The precipitation received during 1978–79 was only 40% of the average precipitation of 1355 mm recorded over the experimental period (1971–88). Soy bean is grown in the monsoon season and yield depends upon the quantity and distribution of rainfall during the growth period. Similarly, the productivity of wheat was also affected during drought years. Small departures from the upward trend in yield in certain seasons appeared to be related to weather aberrations.

While a sharp reduction in yield was observed, in general, with unmanured, N and NP treatments, it was smallest with the NPK, NPKS and NPKSM treatments, indicating that the adverse effect of weather aberrations could be minimized to a great extent, provided the soils were either well fertilized or orginally very fertile. Significant grain yield response of soy bean to applied K was obtained in the drought year of 1978–79 in spite of high native K availability (185 mg/kg) in the soil. It would therefore appear that the diffusion of interlayer $K^+$ from montmorillonite clay seems to be restricted under drier conditions which might become inadequate to meet the K requirement of the crop. Under such situations, significant response to applied K would be obtained (Barber, 1971; Grimme and Van Braunsweig, 1974; Kemmler, 1982; Nambiar, 1985a).

The slow and subtle declining trend in yield of soy bean and wheat with continuous use of high analysis NPK fertilizers was found to be due to emerging S deficiency (Nambiar, 1988). This was quite evident for soy bean and wheat crops after 11–12 annual rotations (Figures 8.10–8.13) with the drawdown in native available S below the critical (5–7 mg/kg) level (Fox, 1976; Fox et al., 1977). The available S (Morgan's-extractable S) decreased to 3–5 mg/kg in the absence of S incorporation after 11–12 annual rotations (Figure 8.18). Application of farmyard manure along with S-containing NPK fertilizers was found to be highly effective not only in correcting S deficiency but also those arising from micronutrients, particularly Zn. The emerging S deficiency with the continuous use of S-free NPK fertilizers was found to have an adverse effect on sustainable production of soy bean, wheat and maize fodder on medium black soil.

## 8.7 SUSTAINABILITY ON LATERITE SOIL AT BHUBANESWAR, ORISSA

Yield trends of monsoon and winter rices, and trends in organic carbon and available soil N, P, K and S as influenced by long-term use of plant nutrients on laterite soil at Bhubaneswar (Orissa) are shown in Figures 8.19–8.26.

### 8.7.1 Yield trend

Grain yield of both monsoon and winter rices indicated a declining trend after 11 cropping cycles, even with recommended doses of mineral NPK fertilizers (Figures 8.19–8.22), which became conspicuous in the subsequent period (Nambiar 1991b; Nambiar, 1992a). However, the abrupt fluctuations in the productivity of monsoon rice in certain seasons seemed to be mainly associated with cloudy weather conditions. The average bright sunshine hours per week during the monsoon seasons (June to September) of 1980, 1983 and 1985 were 5.5, 5.7 and 5.3, respectively whereas they were 7.1, 6.3, 6.4 and 6.4 during

1979, 1981, 1982 and 1984. They varied from 6.1 to 6.4 during 1986–88. The reduced hours of bright sunshine appear to have affected the photosynthesis adversely leading to a sharp reduction in the productivity of monsoon rice. Nonetheless, the gradual and subtle decline in productivity after 11 annual cropping cycles was due to the incipient S deficiency resulting from the continuous use of S-free mineral NPK fertilizers as well as the conversion of the natively available $SO_4^{2-}$ ions into $H_2S$ and $FeS$ under reduced conditions of monsoon rice growing (Mann and Tambane, 1910; Howard, 1925; Nambiar, 1988; Nambiar and Abrol, 1989; Nambiar, 1991b). A diminishing trend in yield of winter rice was noticed with recommended doses of mineral NPK fertilizers after 18 annual cropping cycles. Nevertheless, the productivity of monsoon rice exhibited a steady improvement with integrated use of farmyard manure (M) and mineral NPK fertilizers (Figures 8.19 and 8.20). The residual effect of farmyard manure was also marked on the succeeding winter rice crop (Figure 8.21 and 8.22).

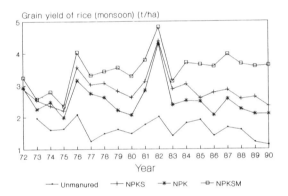

**Figure 8.19** Yield trend of rice (monsoon) on laterite soil in ICAR LTF experiments

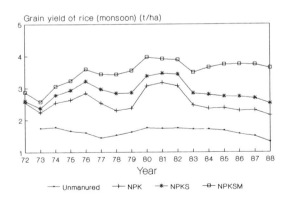

**Figure 8.20** Yield trend (three-year moving average) on laterite soil in ICAR LTF experiments

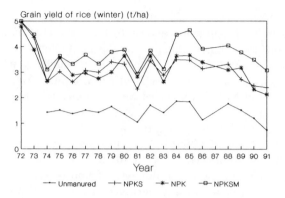

**Figure 8.21** Yield trend of rice (winter) on laterite soil in ICAR LTF experiments

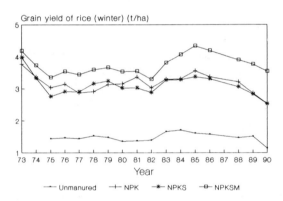

**Figure 8.22** Yield trend (three-year moving average) on laterite soil in ICAR LTF experiments

### 8.7.2 Soil reaction

An overall decrease in soil pH was noted after 17 annual rotations. The maximum reduction was noticed with the NP treatment, the decrease being 0.5 from the initial pH value of 5.5. Integrated use of organic manure and mineral NPKS fertilizers (NPKSM treatment) increased the soil pH by 0.6 over the initial value indicating a moderating effect on soil acidity (Nambiar, 1992b).

### 8.7.3 Soil organic carbon

The trend in soil organic carbon in the rice–rice system showed a small improvement in the case of the unmanured treatment. A similar though larger

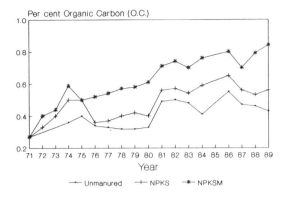

**Figure 8.23** Trend of organic carbon in rice–rice rotation on laterite soil in ICAR LTF experiments

trend also arose with recommended doses of S-containing mineral NPK fertilizers (Nambiar, 1992a). Marked improvement in soil organic carbon was, however, noticed with the NPKSM treatment Figure 8.23).

### 8.7.4 Available soil nitrogen

There was an overall build up in available soil N even at the suboptimal (50%) NPKS dose. Improvement in available soil N amounted to 16.4% over the initial level. Marked improvement in available soil N was recorded with integrated use of organic manure and mineral NPKS fertilizers (NPKSM treatment), the increase being 56.2% over the years 1971–89. However, available soil N decreased by 11.1% over the years 1971–89 on unmanured plots (Nambiar, 1992b).

### 8.7.5 Available soil phosphorus

The initial level of available soil P was maintained with recommended doses of S-containing mineral NPK fertilizers in the initial seven-year period, increased linearly in the subsequent six-year period and levelled off after 14 annual rotations (Figure 8.24). A similar but larger trend was also observed with integrated use of organic manure and mineral NPKS fertilizers (NPKSM treatment). The available soil P decreased markedly for the unmanured treatment in the initial three-year period, steadily declined in the subsequent four-year period and then levelled off (Figure 8.24). The soil is acidic and contains substantial amounts of soluble Fe and Al. The free $Fe_2O_3$ and $Al_2O_3$ in the composite soil sample collected before the layout of the experiment (during 1972) were 0.563% and 0.123%, respectively. Most of the P applied with recommended dose seemed to be fixed in the soil in the initial seven-year period and increased linearly in the subsequent

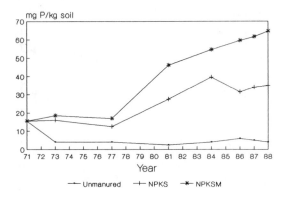

**Figure 8.24** Trend of average P in rice–rice rotation on laterite soil in ICAR LTF experiments

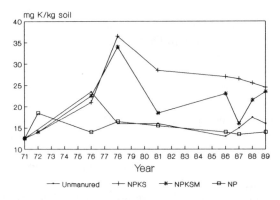

**Figure 8.25** Trend of average K in rice–rice rotation on laterite soil in ICAR LTF experiments

period of eight-years after satisfying the P fixation capacity of the soil. Thereafter, it levelled off indicating an equilibrium state between P uptake by crops and P applied including P release from the native source.

### 8.7.6 Available soil potassium

The available soil K with recommended doses of S-containing mineral NPK fertilizers showed an increasing trend until the eight annual cropping cycle (Figure 8.25). Subsequently it indicated a diminishing trend which was also noticed with the integrated use of organic and mineral NPKS fertilizers (NPKSM treatment). However, reduction in available soil K was greater with the NPKSM treatment than with NPKS (Nambiar, 1992b). This was apparently due to the higher yield harvested under the NPKSM treatment. There was a steadily declining trend for the unmanured treatment, which levelled off after seven annual rotations indicating equilibrium between K uptake and K release from the native source. The non-exchangeable K ($N$ $HNO_3$-extractable K) under NP treatment from 0–45.0 cm depth depleted by $171.5 \pm 29.8$ mg/kg over the years 1971–88 from the initial composite sample value of 393 mg/kg. Similarly, total K from 0–45 cm depth declined by $1100 \pm 169.4$ mg/kg over the years 1971–88 from the initial composite sample value of 5400 mg/kg. It would therefore seem that an appreciable amount of K was transformed from non-exchangeable to available forms (exchangeable K plus water-soluble K), thus maintaining an equilibrium in the absence of K addition over the years 1971–89.

### 8.7.7 Available soil sulphur

The available soil S declined with continuous application of S-free mineral NPK fertilizers over the years 1971–89. However, application of S-containing NPK fertilizers increased the available soil S (Nambiar, 1988) and it increased further with integrated use of organic manure and mineral NPKS fertilizers (Figure 8.26).

### 8.7.8 Discussion of the results at Bhubaneswar

Yield trend of monsoon rice indicated a deterioration in productivity with S-free mineral NPK fertilizers on laterite soil (Bhubaneswar) after three annual cropping cycles (Figures 8.19 and 8.20). Significant reduction was, however, noted after 12 annual cropping cycles as the available S decreased below the critical level (5–7 mg/kg) often reported (Fox, 1976; Fox et al., 1977). The available S dropped from the initial composite sample value of 11.0 mg/kg to $4.5 \pm 0.81$ mg/

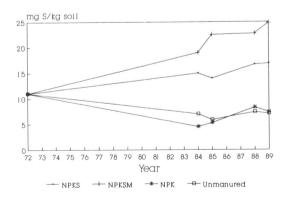

**Figure 8.26** Trend of average S in rice–rice rotation on laterite soil in ICAR LTF experiments

kg after 12 annual rotations (Figure 8.27) with S-free mineral NPK fertilizers and became critical for monsoon rice production (Nambiar, 1988). There was no significant effect on the productivity of winter rice even after 17 annual cropping cycles, although deterioration in productivity was apparent with the 18th annual rotation (Figures 8.11 and 8.12) revealing an incipient S deficiency.

The diminishing trend in monsoon rice productivity was found to be associated with S deficiency due to its conversion to $H_2S$ and FeS under reduced conditions of monsoon rice growing in comparison with winter rice (Mann and Tambane, 1910; Howard, 1925; Nambiar and Abrol, 1989; Nambiar, 1988; Nambiar, 1991b). However, incorporation of S through S-carrying mineral NPK fertilizers checked the deterioration.

A steady improvement in productivity of monsoon rice was noted over the years 1971–89 with the integrated use of farmyard manure and mineral NPKS fertilizers (NPKSM treatment), see Figures 8.19 and 8.20. The residual effect of farmyard manure on the productivity of winter rice was also quite marked (Figures 8.21 and 8.22). The beneficial effect of farmyard manure seemed to be mainly due to a lowering in the activity of Fe and Al in soil solution through the formation of aluminium–organo chelates or iron complexes (Mortensen, 1963; Bartlett and Reigo, 1972; Cabrena and Talibudeen, 1977) as well as from correction of S deficiency in the course of mineralization. Thus, integrating mineral NPKS fertilizers with organic manure was found to be quite promising in maintaining sustainable production of both monsoon and winter rices on laterite soil.

## 8.8 SUSTAINABILITY ON FOOTHILL SOIL AT PANTNAGAR, UTTAR PRADESH

Yield trends of rice and wheat and their average yields over the years 1972–91, yield response to graded doses of NPKS fertilizers, trends in soil pH, organic carbon, available N, P, K and Zn, and also the changes in organic carbon and cation exchange capacity of the soil profile as influenced by long-term use of plant nutrients and intensive cropping on *tarai* (foothill) soil at Pantnagar, Uttar Pradesh, are depicted in Figures 8.27–8.39.

### 8.8.1  Yield trend

Yield trend of rice showed a steady decline in productivity not only with the recommended doses of S-carrying mineral NPK fertilizers but also with the integrated use of farmyard manure and mineral NPKS fertilizers (NPKSM treatment) Figures 8.27 and 8.28. Rice productivity diminished markedly after 15 annual rotations (Nambiar, 1991a; Nambiar, 1992a). The average grain

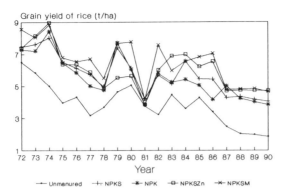

**Figure 8.27** Yield of rice on foothill soil in ICAR LTF experiments

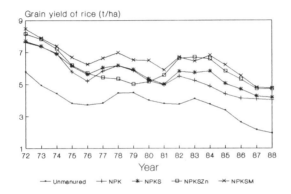

**Figure 8.28** Yield trend (three-year moving average) on foothill soil in ICAR LTF experiments

yield of rice (unhusked) recorded over a period of 15 years (1972–86) dropped from $6.2 \pm 0.123$ t/ha to $4.2 \pm 0.056$ t/ha in the last five-year period (1987–91) with recommended doses of NPKS fertilizers. Similar reductions were also noted with NPKSZn and NPKSM treatments, which were found to be related to the loss of native soil fertility.

The productivity of wheat indicated a declining trend in the absence of Zn and S addition after 11 and 14 annual rotations, respectively, indicating a decline in readily available soil Zn and S below their critical levels (Figures 8.29 and 8.30). The initial wheat productivity was more or less restored with the application of either Zn or farmyard manure along with S-carrying mineral NPK fertilizers.

Linear and quadratic response functions fitted to NPKS at graded doses of 50%, 100% and 150% after 14 and 18 annual rotations for the rice and wheat crops showed a linear trend for wheat indicating that considerable scope still existed for improving the productivity of wheat with increases in NPK doses (Figure 8.31).

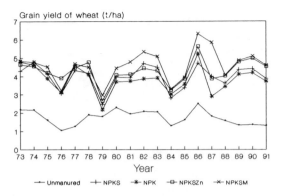

**Figure 8.29**  Yield trend of wheat on foothill soil in ICAR LTF experiments

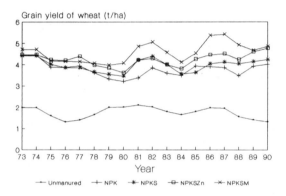

**Figure 8.30**  Yield trend (three-year moving average) on foothill soil in ICAR LTF experiments

Similarly, the response and profit maximizing NPK doses increased considerably for rice over the initially recommended doses (Table 8.4 and Figure 8.32), thus confirming the loss of inherent soil fertility and productivity.

### 8.8.2  Soil reaction

Soil pH increased by 0.5 over the initial value of 7.3 with the integrated use of farmyard manure and S-containing mineral NPK fertilizers (NPKSM treatment) over a four-year period but evened out subsequently. However, the soil pH steadily increased with the NP treatment over an eight-year period and levelled off in the subsequent four-year period which was then followed by an increasing trend. The maximum increase in soil pH was noticed with the NP treatment, the increase being 0.9 (Nambiar, 1992a). A similar trend was noticed with the

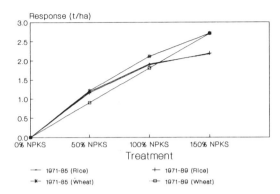

**Figure 8.31** Grain yield response curve of rice and wheat on foothill soil in ICAR LTF experiments

unmanured treatment but the increase after 12 annual rotations was more than with the NPKSM treatment but less than with the NP treatment (Figure 8.32).

The general increase in soil pH seems to have resulted from soil K mining and its partial replacement by $Na^+$ ions on the soil exchange complex due to inadequate K application over the years (1972–90). The effect was more pronounced for the NP treatment, where the soil pH increased by 0.9 implying a relatively greater entry of $Na^+$ ions on the exchange complex of the soil. Consequently, the availability of Zn become critical after 12 annual rotations leading to a decline in productivity.

### 8.8.3 Soil organic carbon

The declining trend in soil organic carbon was noticed in the case of the unmanured treatment after nine annual rotations. However, the diminishing trend in soil organic carbon commenced with the NPKS and NPKSM treatments only after 12 annual rotations (Figure 8.33). Mineralization of organic matter was quite rapid in the subsequent two-year period and thereafter it levelled off. However, loss of soil organic carbon was least with the NPKSM treatment (Nambiar, 1992a). Application of mineral NPK fertilizers seemed to have maintained the initial C:N ratio of the soil for about 12 years but appeared to have failed to maintain it subsequently due to heavy withdrawal of N from the soils far in excess of the applied N which resulted in accelerated mineralization and consequent loss of soil organic matter. Incorporation of farmyard manure (along with S-containing mineral NPK fertilizers) more or less maintained the initial level of organic carbon but showed a gradual decline with heavy N withdrawal from the soil.

The overall reduction in soil organic carbon under the rice–wheat–cowpea fodder system appears to be related to the mineralization of organic matter due to

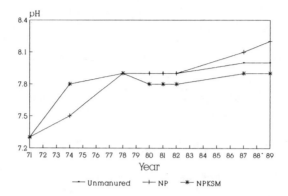

**Figure 8.32** Trend of soil pH in rice–wheat–cowpea (fodder) rotation on foothill soil in ICAR LTF experiments

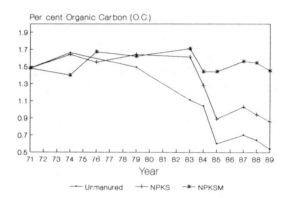

**Figure 8.33** Trend of organic carbon in rice–wheat–cowpea (fodder) rotation on foothill soil in ICAR LTF experiments

intensive cropping on the one hand, and differing growing conditions of rice and wheat involving alternate wetting and drying, thereby inducing accelerated mineralization of soil organic matter, on the other (Nambiar, 1992a).

### 8.8.4 Distribution of organic carbon in the soil profile

The distribution of organic carbon in the soil profile, studied up to a depth of 90 cm after 15 annual rotations, showed that the organic carbon declined throughout the profile with both the NPKS and unmanured treatments. The initial composite soil organic carbon content of the plough layer (0–15 cm) dropped from 1.6% to $1.03 \pm 0.019\%$ with the NPKS treatment. Similarly, it declined to $0.70 \pm 0.019\%$ in the plough layer (0–15 cm depth) with the un-

**Table 8.4** Response and profit maximizing doses (kg/ha) in ICAR Long Term Fertilizer Experiments

| Location/ Crops | Recommended dose (kg/ha) at 100 NPK dose | | | Response maximizing dose (kg/ha) | | | Response (kg/ha) at response maximizing dose | Profit maximizing dose (kg/ha) | | | Response (kg/ha) at profit maximizing dose |
|---|---|---|---|---|---|---|---|---|---|---|---|
| | N | $P_2O_5$ | $K_2O$ | N | $P_2O_5$ | $K_2O$ | | N | $P_2O_5$ | $K_2O$ | |
| **Pantnagar** | | | | | | | | | | | |
| Rice | | | | | | | | | | | |
| 1971–85 | 120 | 60 | 45 | 184 | 92 | 69 | 2173 | 135 | 68 | 51 | 2021 |
| 1971–89 | 120 | 60 | 45 | 193 | 97 | 72 | 2204 | 150 | 75 | 56 | 2096 |
| Wheat | | | | | | | | | | | |
| 1971–85 | 120 | 60 | 40 | † | † | | — | † | † | | — |
| 1971–89 | 120 | 60 | 40 | † | † | | — | † | † | | — |

† Linear trend

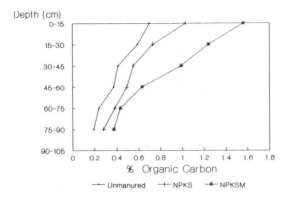

**Figure 8.34** Profile organic carbon dynamics in rice–wheat–cowpea rotation (after 15 cycles) on foothill soil in ICAR LTF experiments

manured treatment, indicating a negative carbon balance and consequent loss of soil fertility and productivity (Figure 8.34).

### 8.8.5 Cation exchange capacity

The change in cation exchange capacity (CEC) in the soil profile was also studied up to a depth of 90 cm after 15 annual rotations.

The cation exchange capacity of the initial composite sample of the plough layer (0–15 cm depth) was 20 c mol ($p^+$)/kg which was more or less maintained with the NPKS treatment even after 15 annual cropping cycles. It showed an increase of $2.8 \pm 0.082$ c mol ($p^+$)/kg with the NPKSM treatment. However, the CEC declined by $3.4 \pm 0.082$ c mol ($p^+$)/kg in the plough layer with the unmanured treatment, indicating a reduction in the nutrient retention capacity of the soil (Figure 8.35).

### 8.8.6 Available soil nitrogen

The trend in available soil N showed a sharp decline with the NPKS and NPKSM treatments after 10 and 13 annual rotations, respectively (Figure 8.36). Reduction in available soil N was linear with the unmanured treatment up to the 10th annual rotation and subsequently attained equilibrium. Interestingly, the trend in available soil N after 17 annual rotations was more or less similar with the unmanured, NPKS and NPKSM treatments, indicating an equilibrium between N uptake and N applied (including the release of N from the native source). It would therefore appear that the initially recommended doses of mineral NPK fertilizers had become suboptimal for maintaining sustainable crop production.

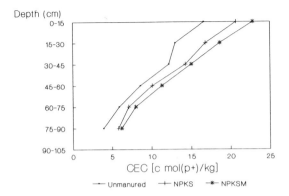

**Figure 8.35** Profile CEC dynamics in rice–wheat–cowpea rotation (after 15 cycles) on foothill soil in ICAR LTF experiments

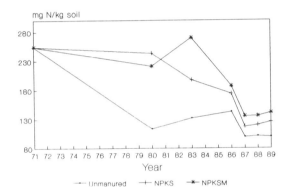

**Figure 8.36** Trend of average N in rice–wheat–cowpea (fodder) rotation on foothill soil in ICAR LTF experiments

This was also revealed in the reduction of N uptake after 15 annual rotations (Nambiar, 1992a).

### 8.8.7 Available soil phosphorus

The trend in available soil P showed an overall increase with the unmanured, NPKS and NPKSM treatments in the initial three-year period. Subsequently, a sharp reduction in available P was noticed with the unmanured treatment. This flattened off after eight annual rotations. Sharp reduction in available soil P was associated with the uptake of P by crops from the native source when no fertilizer P was applied (Figure 8.37). However, there was no reduction in available soil P with recommended doses of mineral NPKS fertilizers until the 17th annual

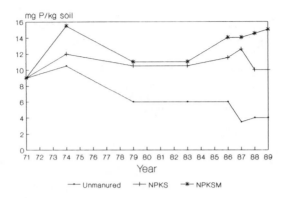

**Figure 8.37** Trend of average rice–wheat–cowpea (fodder) rotation on foothill soil in ICAR LTF experiments

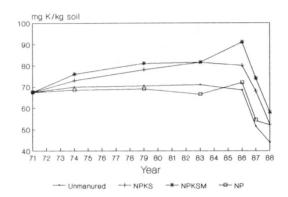

**Figure 8.38** Trend of average K in rice–wheat–cowpea (fodder) rotation on foothill soil in ICAR LTF experiments

rotation. Subsequently, there was a small reduction in available soil P indicating lack of balance between P uptake by crops and P applied (including P released from the soil). The available soil P with the NPKSM treatment increased in the initial three-year period, declined in the subsequent five-year period and levelled off in the following four-year period. Thereafter, it showed an increasing trend (Figure 8.37). The rise in available P in the last five-year period (1987–91) seems to be connected with low P uptake by rice due to decreased rice yield (Nambiar, 1992a).

### 8.8.8 Available soil potassium

There was a marginal improvement in available soil K ($N$ $NH_4OAc$-extractable K) with the NPKS and NPKSM treatments until the 16th annual rotation

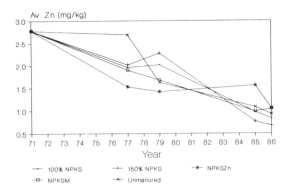

**Figure 8.39** Trend of average soil Zn on foothill soil in ICAR LTF experiments

(Figure 8.38). The subsequent sharp decline was also revealed in the productivity of rice and wheat crops (Nambiar, 1992a). The annual addition of K was only 70 kg/ha with recommended doses of mineral NPKS fertilizers as against the average annual K uptake of about 286 kg/ha, indicating considerable soil mining of readily available soil K leading to loss of native soil K fertility and productivity.

### 8.8.9 Available soil sulphur

The available soil S ($CaCl_2$-extractable S) dropped to around 9.1 mg/kg with the continuous use of S-free NPK fertilizers over the years 1971–89. At this level of available S, rice productivity decreased. Similar reductions in available S were also observed with the unmanured and N treatments over the years (1971–89), the available S being about 13.2 and 16.5 mg/kg, respectively. The lower reductions with these treatments compared with S-free NPK treatment were found to be due to lower yields resulting from imbalanced use of plant nutrients.

### 8.8.10 Available soil zinc

Available soil Zn (DTPA-extractable Zn) declined from the initial composite value of 2.6 mg/kg to $0.83 \pm 0.035$ mg/kg (Nambiar, 1989) over a period of 15 years (1971–86), see Figure 8.39. It became limiting for rice and wheat production when the available soil Zn dropped to $0.97 \pm 0.113$ mg/kg (Nambiar, 1986). The productivity of rice and wheat showed significant improvement with foliar application of $ZnSO_4 \cdot 7H_2O$ (5.0 kg/ha) to each crop. Similarly, incorporation of farmyard manure (along with S-containing mineral NPK fertilizers) maintained the productivity of both rice and wheat crops (Nambiar and Abrol, 1989; Nambiar, 1991b; Nambiar, 1992a).

Other micronutrients like Fe, Mn and Cu had no significant effect on the productivity of rice and wheat crops since the levels of these nutients were still higher than their critical levels.

### 8.8.11 Discussion of the results of Pantnagar

The productivity of rice on *tarai* (foothill) soil at Pantnagar, Uttar Pradesh declined sharply after 15 annual rotations in all the treatments (Nambiar, 1989; Nambiar, 1992a). The average grain yield over a period of 15 years (1972–86) declined from $6.2 \pm 0.123$ t/ha to $4.2 \pm 0.056$ t/ha in the last five-year period (1987–91) with recommended doses of mineral NPKS fertilizers. Similar reductions were also noted with 150% NPKS, NPKSZn and NPKSM treatments. The rice grain productivity with 150% NPKS treatment declined by $0.6 \pm 0.015$ t/ha in comparison with 100% NPKS treatment after 18 annual rotations (1989–90). The reduction was found to be associated with acute Zn deficiency as a result of enhanced Zn uptake with increase in fertilizer NPK doses, which led to a decrease in available soil Zn (Figure 8.39) in the absence of its application over the years (1972–91).

The average rice productivity of $4.2 \pm 0.056$ t/ha (unhusked) recorded in the last five-year period (1987–91) could, however, be maintained either with the application of Zn or farmyard manure (15 t/ha/yr) along with recommended doses of S-containing mineral NPK fertilizers. The increase in rice grain productivity with foliar application of Zn during 1990 was $1.3 \pm 0.015$ t/ha over the 150% recommended doses of mineral NPKS fertilizers. Nevertheless, the overall decline in rice productivity was found to be related to the reduction in inherent soil fertility as a consequence of intensive cropping on the one hand, and inadequate application of plant nutrients especially NPK on the other.

The quadratic response functions fitted to NPKS at graded doses also showed that response and profit maximising doses increased considerably for rice over the initially recommended doses (Table 8.4 and Figure 8.31) further confirming the loss of inherent soil fertility. The decline in rice productivity was further accentuated by the deficiencies arising from secondary nutrients like S and micronutrients like Zn. The availability of Zn was also adversely affected due to the rise in soil pH (Figure 8.32). Loss of inherent soil fertility was also evidenced by sharp reduction in soil organic matter, cation exchange capacity and available soil N, K, S and Zn. Thus, the falling trend in rice productivity could be ascribed to the compounded effects of various fertility parameters involved in soil fertility and productivity management.

The initial productivity of wheat was, however, maintained with the recommended doses of mineral NPKS fertilizers throughout the experimental period (1972–91). Increase in doses of NPKS fertilizers by 50% over the 100% NPKS treatment increased the average grain productivity of wheat by $0.56 \pm 0.072$ t/ha over the 1972–89 period in comparison with the 100% NPKS treatment.

However, wheat grain productivity with the 150% NPKS treatment was $0.63 \pm 0.013$ t/ha less than with the 100% NPKS treatment during 1990–91, and a similar reduction occured in rice productivity. This was also found to be associated with acute Zn deficiency (Nambiar, 1992b). Foliar application of Zn (5 kg $ZnSO_4 \cdot 7H_2O$/ha) increased the grain productivity by $1.28 \pm 0.013$ t/ha over the 100% NPKS treatment. Incorporation of farmyard manure (along with S-carrying NPK fertilizers) maintained a grain productivity of wheat similar to that of the Zn treatment. Thus, the productivity of wheat could be maintained through correction of Zn deficiency either with the application of Zn or farmyard manure along with recommended mineral NPKS fertilizers. Considerable scope still exists for enhancing wheat productivity with increases in NPK fertilizer doses. This was also confirmed by the quadratic response functions fitted to NPKS at graded doses which showed a linear trend (Figure 8.31) indicating the need for larger NPK dressing for maintaining higher wheat productivity on *tarai* (foothill) soil.

## 8.9 CONCLUSION

The biological sustainability of maize in maize–wheat–cowpea fodder rotation on alluvial sandy soil at Ludhiana (Punjab) was significantly affected by the deficiency of Zn after 10 annual rotations when the available soil Zn decreased from 1.1 to around 0.68 mg/kg. The yield sustainability could, however, be maintained with the application of Zn (50 kg $ZnSO_4 \cdot 7H_2O$/ha) once in 3–8 years or farmyard manure at 15 t/ha/yr to the maize crop along with recommended doses of mineral NPKS fertilizers. The sustainability of wheat was, nevertheless, maintained by the recommended doses of mineral NPKS fertilizers. Thus, the highest productivity of maize was noticed with the integrated use of farmyard manure and mineral NPKS fertilizers.

The sustainability of soy bean, wheat and fodder maize was found to be affected on medium black soil at Jabalpur (Madhya Pradesh) after 11–12 annual rotations by the depletion of available S (Morgan's-extractable S) below the critical level (5–7 mg/kg). The productivity of soy bean, wheat and fodder maize was restored by the application of S-containing NPK fertilizers or farmyard manure along with recommended doses of S-containing mineral NPK fertilizers.

Similarly, the sustainable productivity of monsoon and winter rices was affected by the deficiency of S on laterite soil at Bhubaneswar (Orissa) after 12 and 17 annual rotations, respectively. The biological sustainability of the rice–rice cropping system was also affected by accumulation of Fe and Al to toxic levels interfering with the uptake of P and micronutrients. It could, however, be counteracted by lime amendment or incorporation of farmyard manure along with mineral NPKS fertilizers. Deficiency of S could be corrected by the application of S-containing NPK fertilizers. Thus, the biological sustainability of the rice–rice system could be maintained by the integrated use of farmyard manure and S-containing mineral NPK fertilizers or lime amendment followed by NPK dressings.

The biological sustainability of rice in the rice–wheat–cowpea fodder system was severely affected on *tarai* (foothill) soil at Pantnagar (Uttar Pradesh) after 15 annual rotations. The overall decline in rice productivity was found to be associated with the reduction in inherent soil fertility as a consequence of intensive cropping and inadequate application of plant nutrients especially NPK. The decline in rice productivity was further accentuated by the deficiencies arising from S and Zn. Loss of native soil fertility was also exhibited through sharp reduction in soil organic matter, cation exchange capacity and available soil N and K. The diminishing trend in rice productivity could be ascribed to the compounded effects of various fertility parameters involved in the decline of soil quality.

Nonetheless, the initial productivity of wheat was maintained with the recommended doses of S-containing NPK fertilizers. Reduction in productivity of wheat with higher NPKS (150% NPKS) doses was found to be affected by acute Zn deficiency. However, deficiency of Zn could be corrected either by the application of Zn or farmyard manure along with S-containing mineral NPK fertilizers.

Thus, periodic monitoring of nutrients would help reveal the emerging multinutrient deficiencies arising from intensive cropping that often impair sustainable productivity regardless of an adequate quantity of all other growth factors at optimal levels. It would help track deficiencies of various nutrients as and when they occur. It is quite evident from the results of the All India Coordinated Long Term Fertilizer Experiments that sustainable productivity of a high order cannot be achieved in intensive cropping systems with the use of organic manure alone nor with the exclusive use of mineral NPK fertilizers, due to their inherent limitations. However, integrating organic sources with adequate S-containing mineral NPK fertilizers has been found to be quite promising in maintaining sustainable production while enhancing soil fertility through complementary effects. The influence of organic manures in combination with minearl NPKS fertilizers has been all the more revealing because, as a soil ameliorant, they avoid the adverse soil ecology of acid soils, leading to improved soil fertility and productivity.

### 8.9.1 Gaps and future research needs

The nutrient turnover in intensive cropping systems is so large that even the inherent fertility of well fertilised soils, over the long-term, is found to be in decline and productivity is adversely affected. This is especially true in the case of rice–rice and rice–wheat systems. Maintaining sustainable production of a high order without decline in inherent soil fertility is a colossal task. Nonetheless, the goal of sustainable crop production can be achieved to a great extent by scientific

# Conclusion

management of production practices. The following research needs to be taken up as a priority to achieve this goal.

(i) Nutrient management on a cropping system basis instead of on an individual crop basis.
(ii) Identification and monitoring of nutrients including major, secondary and micronutrient elements for forewarning of nutrient stresses that limit crop production.
(iii) Identifying and monitoring accumulation of heavy metals such as cadmium under long-term phosphate application.
(iv) Determining the fate of applied N over the long term including ground water pollution of $NO_3^-$ using $^{15}N$ techniques.
(v) Management of residual P build up and enhancing P use efficiency.
(vi) Monitoring pesticide build up in soils and plants under long-term fertilizer application including ground water pollution.
(vii) Monitoring pests and disease complexes under long-term fertilizer use.
(viii) Studying microbial population dynamics, C and N turnover and their role in nutrient transformation in multiple cropping systems.
(ix) Recycling of organic residues including crop residues in an integrated manner with inorganic fertilizers for complementary fertilization.
(x) Developing the nutrient balance sheet under multiple cropping systems, based on the contributions of nutrients from fertilizers, lime, manure, additions from legumes, release of nutrients from primary minerals or reserves of fixed nutrients applied in the past and nutrients removed in grain and straw, fodder and fruits.
(xi) Finally, modelling production systems to enable the prediction of yield potentials or realistic yield targets, based on climate, known physical and chemical properties of the soils and responses to fertilization and amendments as well as quantification of incoming and outgoing nutrient fluxes among the components of yield and their effect on total crop yield that can be validated over a wider range of environmental conditions over the years.

Complete information about these aspects would help develop strategies not only for nutrient management but also production systems management as a whole so that we might maintain sustained higher productivity without compromising soil fertility.

# 9

# Sustainability—the Rothamsted Experience

V. Barnett[†], A. E. Johnston[†], S. Landau[†], R. W. Payne[†],
S. J. Welham[†] and A. I. Rayner[‡]

[†]*Rothamsted Experimental Station, UK*
[‡]*University of Nottingham, UK*

## 9.1 INTRODUCTION

In all developed countries we hear widespread and vociferous demands for the adoption of agricultural systems which are 'sustainable'. The stimulus comes from various sources in the environmental, economic, legal, social and political structure. It has received increasing attention in recent official pronouncements of EC agricultural policy. For example, in the important green paper on reform of the Common Agricultural Policy—the so-called MacSharry Plan (after the EC Agricultural Commissioner, Ray MacSharry)—it is recognised that without fundamental reform 'the in-built incentive to greater intensity and further production, provided by the present mechanisms, puts the environment at increasing risk' (EC Commission 1991, page 3). Moreover, 'more specific incentives towards environmentally friendly farming should be available [and] there should be greater recognition of the dual role of the farmer in producing food and managing the countryside' (ibid).

Undoubtedly, it is a laudable aim to seek an agricultural policy which is responsive to environmental needs, and receptive to economic, legal and social demands, and yet it is not entirely obvious what is meant by *sustainable agriculture*. It is variously described and defined depending on the stand-point of those who are discussing it. At its most fundamental level it implies a mechanism by which *the quality and quantity of agricultural yield can be maintained year*

---

*Agricultural Sustainability: Economic, Environmental and Statistical Considerations.*
Edited by V. Barnett, R. Payne and R. Steiner © 1995 John Wiley & Sons Ltd

after year without degradation of the soil, environmental contamination, disruption of habitats for flora and fauna, pollution of water courses, etc. Whilst this is generally regarded as what is meant by sustainability, differences begin to arise when seeking to define quality, quantity and maintenance.

As an extreme example, the total replenishment each year of the top layer of soil on a farm by fully nourished and enriched new soil will presumably ensure that we continue to produce 'quality and quantity'. Whether a particular prospect is feasible in economic terms is sometimes not considered. And yet no agricultural system can operate in isolation, in disregard of utility and cost. It therefore seems essential that we add an economic component to any definition of sustainability, i.e. *regarding sustainability as the maintenance of quality and quantity when measured in regard to the difference between output values and input costs*.

However, quantification remains a major research issue. In a limited sense, the ability of an agricultural system to maintain a consistent flow of marketable output over time is an indicator of sustainability. But this ignores the mix and quantities of inputs used in conjunction with land, the basic natural capital. It also fails to account for 'externalities' or environmental degradation, such as the *environmental costs* of nitrate leaching into a neighbouring water course, or of pesticide transfer and carry-over. So for operational purposes, sustainability might be assessed in relation to an index of 'total factor productivity' with due allowance for externalities such that a 'sustainable system has a non-negative trend in total factor social productivity over the period of concern' (Lynam and Herdt, 1989). Of course, quantifying movements in total factor social productivity over time is no easy task because of information requirements. A time series database on quantities of outputs and inputs must be compiled; weights for combining outputs (inputs) into a single index of production (resource use) need to be determined; index number issues have to be addressed; and off-site costs and benefits evaluated.

This study seeks to measure the relationship between the output value and the input costs of an agricultural system for the purpose of determining whether, in the long term, the relative difference is positive and is being maintained.

The resources of the Rothamsted long-term experiments (dating back 150 years in some instances) provide a valuable basis for an investigation of the sustainability of crop production systems by means of a total factor social productivity measure. These 'Classical experiments' provide a unique source of information and allow meaningful comparisons between crop systems, where one system (or experiment) being studied may clearly be sustainable whilst another may not.

A brief review is necessary at this stage of the Classical experiments at Rothamsted and of the role they have played in agricultural research.

Rothamsted Experimental Station is the oldest existing agricultural research station in the world, dating its foundation to 1843. John Bennet Lawes, who

## Introduction

became a successful scientist, businessman and philanthropist, had, like others, found that treating bones with sulphuric acid enhanced the solubility of the phosphate, which when applied to arable crops, improved their yields. At the same time, he felt it essential that scientific experimentation on crop nutrition should be carried out: hence Rothamsted was established (see Stevenson, 1989; Johnston, 1994; Johnston and Powlson, 1993).

The earliest activities consisted of setting out some large areas of land on which to conduct experiments to examine over time the way in which applied fertilizers influenced crop yield, diversity and soil conditions (with coincident monitoring of meteorological data). An interesting feature of these 'quasi-designed statistical experiments' was that they *predated* both the study of designed experiments and the formal development of statistical methods *by more than 50 years*.

Two of the principal Classical experiments ('Classicals') at Rothamsted which we considered were started in the 1840s and 1850s. They are *Park Grass 1856*, (2.8 hectares, grassland), *Broadbalk 1843* (4.6 hectares, winter wheat).

Although *minor* variations in the treatments, husbandry and crops have taken place over the 150 years (and some plots have been divided to allow new treatments), there has been an impressive degree of consistency over this long period: the treatment combinations have remained largely unaltered on Broadbalk for 140 years. Again, although the treatment designs are incomplete, there is still a strong manifestation of what we now think of as *crossed-factor experimental designs*, sometimes with *split-plot* components, with most factors at two levels (absence or presence) and some at four or more. Park Grass provides a good example of the experimental aims, being concerned with the effects of inorganic and organic fertilizers, and later of lime ($CaCO_3$), on grass for hay production. The latest of the Classicals, Park Grass, is now essentially a split-plot factorial experiment with a 6-level factor for N, 2-level factors for P, K, Na, Mg and farmyard manure, and lime employed as the subplot treatment since 1903. With initially only 20 plots (of which many were halved in 1903 and then split into four subplots since 1965), we are far from a complete design, but the design logic is self-evident, anticipating the balanced complete designs which came into prominence with supporting methodology so much later in industrial, argicultural (and other) studies from the 1920s onwards. Some interesting effects are reported by Anon. (1991):

> The Park Grass experiment, laid down in 1856, is much the oldest on grassland in Great Britain. The field had been pasture for at least a century when the experiment began. It demonstrates in a unique way how continued manuring with different fertilizers affects both the botanical composition and the yield of a mixed population of grasses, clovers and other herbs. After more than 130 years, the boundaries of the plots are still clearly defined; the transition between adjacent treatments occupies 30 cm or less, showing that there is little sideways movement of nutrients in flat undisturbed soil.

The 'Classicals' provide a unique opportunity to carry out multivariate environmental statistical analysis on long-term data of great complexity, even in terms of present sophisticated statistical methodology. They provide the rich source of material on which we conducted our part of this Rockefeller project on sustainability. The special advantages of using data from the 'Classicals' include

(i) the unequalled length of the time-period—although this posed some major problems in discovering some of the quantitative and economic data (especially data not normally reported for experiments);
(ii) the wide variety of experiment type in terms of crops, farming procedures, treatments, etc.—which enabled us to choose examples revealing different degrees of sustainability or degradation.

Data from Park Grass and Broadback were used in this study. We also used data from the Woburn Continuous Wheat experiment, which ran from 1877 to 1926, and was stopped because yields dropped to very low levels.

## 9.2 OBJECTIVES OF THE ROTHAMSTED STUDY

*The overall aim of this project was to assess the sustainability of an agricultural system by relating it to different measures of total (social) factor productivity (TSFP).* Clearly, in order to feel confident in the use of any such measures, it is necessary to know that they will behave as expected, reflecting the sustainability or non-sustainability of the system. The need to calibrate the different possible measurements was a major consideration in our selection of experiments for analysis. Thus, our choice of experimental plots for inclusion in the study was designed to range from those likely to be sustainable (Park Grass) to unsustainable (Woburn Continuous Wheat) in order to see how the different possible indexes would reflect these differing situations. Broadbalk was then included as an example of continuous cereal growing; the rotational wheat being similar to current practice.

In each experiment, the treatments applied to each plot had remained almost constant over time, and consisted mainly of fertilizer and lime applications. On Broadbalk, there were also different cropping regimes. A range of plots was chosen from each experiment to cover both realistic and extreme farm conditions; these are summarized below in Table 9.1. Details of experimental treatments can be found in Tables 9.2–9.4.

Given the long-term nature of these experiments, we would expect any numerical measures to verify the sustainable nature of the Park Grass experiment and the disastrously unstable nature of the Woburn Continuous Wheat cropping system, as well as allowing us to assess the performance of a fairly typical cereal production system on Broadbalk. These experiments are discussed in more detail below.

## Objectives of the Rothamsted study

**Table 9.1** Experimental treatments on plots chosen for the sustainability study

| Treatment | N (level) | FYM | P | K | Na | Mg | Lime | Pesticides |
|---|---|---|---|---|---|---|---|---|
| **Park Grass** | | | | | | | | |
| Plot 3(d) | – | – | – | – | – | – | – | – |
| Plot 11/1(a) | 3 | – | × | × | × | × | × | – |
| Plot 11/1(d) | 3 | – | × | × | × | × | – | – |
| Plot 14/2(d) | 2 | – | × | × | × | × | – | – |
| **Woburn Continuous Wheat (plot 8a split in 1905 to give 8a and 8aa)** | | | | | | | | |
| Plot 8a | 4 | – | × | × | × | × | – | – |
| Plot 8aa | 4 | – | × | × | × | × | × | – |
| **Broadbalk Continuous Wheat and Wheat in Rotation** | | | | | | | | |
| Plot 21 | 2 | × | – | – | – | – | × | × |
| Plot 22 | – | × | – | – | – | – | × | × |
| Plot 3 | – | – | – | – | – | – | × | × |
| Plot 8 | 3 | – | × | × | × | × | × | × |
| Plot 9 | 4 | – | × | × | × | × | × | × |

Note: × indicates treatment applied.

### 9.2.1 Park Grass

Since sustainability is largely undefined, particularly in relation to externalities, finding a sustainable system against which to judge TSFP measures is not an easy task. However, at least some of the plots on the Park Grass experiment (1856 onwards) at Rothamsted do appear to meet an intuitive definition of sustainability: from the same inputs each year (a constant level of fertilizer application), the yield of hay on some of the plots has remained reasonably stable. Obviously, this is the case for many agricultural systems, but externalities must then be taken into account. On the Park Grass experiment, no pesticides have been used. Furthermore, we can restrict our analysis to the plots where a moderate amount of fertilizer has been used, and so externality effects due to fertilizer application might be negligible. The different lime ($CaCO_3$) and fertilizer treatments clearly affect the botanical composition and diversity of some plots, and this should be assessed as an external effect of the treatment. However, plots also exist where the original soil pH has been approximately maintained and the botanical composition is approximately stable. Plots 14/2(d) (fertilizer application N2*PKNaMg, see Table 9.2 for details) and 11/1(a) (N3PKNaMg, limed since 1903 and, since 1965, maintained at pH 7) both fall into this category.

For comparison, we also took two plots with management systems which we considered potentially non-sustainable. Plot 11/1(d) (N3PKNaMg), where no

**Table 9.2** Fertilizer application on Park Grass.

| Code | | Annual Application | Composition |
|---|---|---|---|
| N2* | 1856–1991 | 617 kg/ha nitrate of soda | 15.5% N |
| N3 | 1856–58 | 818 kg/ha mixed ammonium salts[a] | 23.5% N |
| | 1859–61 | 409 kg/ha mixed ammonium salts | 23.5% N |
| | 1862–81 | 818 kg/ha mixed ammonium salts | 23.5% N |
| | 1882–1916 | 614 kg/ha mixed ammonium salts | 23.5% N |
| | 1917–91 | 686 kg/ha sulphate of ammonia | 21% N |
| P | 1856–88 | 392 kg/ha single-superphosphate[b] | 18% $P_2O_2$ |
| | 1889–96 | 407 kg/ha single-superphosphate | 18% $P_2O_5$ |
| | 1897–1902 | 448 kg/ha basic slag | – |
| | 1903–73 | 407 kg/ha single-superphosphate | 18% $P_2O_5$ |
| | 1974 | 108 kg/ha triple-superphosphate | 47% $P_2O_5$ |
| | 1975–86 | 407 kg/ha single-superphosphate | 18% $P_2O_5$ |
| | 1987–91 | 108 kg/ha triple-superphosphate | 47% $P_2O_5$ |
| K | 1856–78 | 336 kg/ha sulphate of potash | 49% $K_2O$ |
| | 1879–1916 | 560 kg/ha sulphate of potash | 49% $K_2O$ |
| | 1919–91 | 560 kg/ha sulphate of potash | 49% $K_2O$ |
| Na | 1856–63 | 224 kg/ha sulphate of soda | 14% NA |
| | 1864–1991 | 112 kg/ha sulphate of soda | 14% NA |
| Mg | 1856–1991 | 112 kg/ha sulphate of magnesia | 10% Mg |

Note: See Table 9.4.

lime has been applied and the soil pH has now declined to 3.5, is dominated by one grass species which is high yielding but visually unattractive. Plot 3(d), where no fertilizer and no lime have been applied—resulting in maximum species diversity and visual attractiveness—has very low yields which could make it uneconomic in farming terms.

### 9.2.2 Woburn continuous wheat

Although it is difficult to ascertain the long-term sustainability of any given system, it can be much easier to identify an unsustainable system. The continuous wheat experiments carried out by the Royal Agricultural Society of England at the Woburn experimental farm from 1877 to 1926 provide a good example of this. On plot 8a of the experiment, winter wheat was grown continuously with the application of ammonium sulphate as a nitrogen fertilizer (for full details see Table 9.3) which, since no lime was applied, caused a rapid increase in the acidity of the soil and a similarly rapid decrease in yield. In 1905 the plot was

### Objectives of the Rothamsted study

**Table 9.3** Fertilizer application on Woburn Continuous Wheat

| Code | | Annual Application | Composition |
|---|---|---|---|
| N4 | 1877–81 | 400 kg/ha mixed ammonium salts[a] | 23.5% N |
|    | 1882–1905 | 400 kg/ha mixed ammonium salts, alternate years | 23.5% N |
|    | 1906–26 | 220 kg/ha sulphate of ammonia, alternate years | 21% N |
| P  | 1877–88 | 392 kg/ha single-superphosphate[b] | 18% $P_2O_5$ |
|    | 1889–1906 | 586 kg/ha single-superphosphate | 18% $P_2O_5$ |
|    | 1907–26 | 478 kg/ha single-superphosphate | 14% $P_2O_5$ |
| K  | 1877–1906 | 225 kg/ha sulphate of potash | 49% $K_2O$ |
|    | 1907–26 | 61 kg/ha sulphate of potash | 49% $K_2O$ |
| Na | 1877–1926 | 112 kg/ha sulphate of soda | 14% Mg |
| Mg | 1877–1926 | 112 kg/ha sulphate of magnesia | 10% Mg |

Note: See Table 9.4

split, and lime applied to one half (plot 8aa) in an attempt to remedy the situation. This is an extreme example, but any reasonable measure of sustainability should clearly identify the unsustainability of this system, and so this experiment acts as a good bench-mark.

#### 9.2.3 Broadbalk

The third experiment considered as part of this study was Broadbalk (1843 onwards); see Dyke *et al.* (1982). Since 1852 wheat has been grown on all or part of the experiment continuously and, for some periods, in rotation. Of the continuous wheat plots, Broadbalk is somewhat atypical of current arable farm practice in the UK but, with the management practices adopted, it is a remarkable example of a sustained agricultural production system. Various levels of fertilizer application, including farmyard manure (FYM) have been used on the different plots, with the addition of herbicides and pesticides from the 1960s onwards. This gave us the chance to examine the effects of externalities in this farming system.

Broadbalk is arranged as a two-way classification: different fertilizer treatments are applied to (nowadays) 19 plots, which are in turn divided into 10 sections to test different cropping systems. The Broadbalk plots chosen were: 21 (N2FYM—medium fertilizer nitrogen level plus farmyard manure), 22 (FYM only), 3 (no fertilizer), 8 (N3PKNaMg—moderate nitrogen level) and 9 (N4PKNaMg—high nitrogen level). The treatments on plots 22, 3 and 8 have been applied since the start of the experiment, while the treatments on plots 21 and 9 were introduced in 1968. A detailed description of fertilizer applications is given in Table 9.4.

Originally, each plot of Broadbalk was treated as a single unit. In 1926 each plot was split into 5 sections (I–V) to allow fallowing as a means of weed control. From 1931–1967 a regular system of fallowing took place: the five sections were fallowed in turn, so that each section grew four wheat crops followed by one year's fallow with summer cultivation to keep down weeds. This scheme can only be regarded as a regular rotation since 1935. Further changes took place

**Table 9.4** Fertilizer application on Broadbalk Continuous Wheat and Wheat in Rotation

| Code | | Annual Application | Composition |
|---|---|---|---|
| N2 | 1968–85 | 457 kg/ha 'Nitro Chalk'[c] | 21% N |
|    | 1986–89 | 278 kg/ha 'Nitram' | 34.5% N |
| N3 | 1852–1916 | 610 kg/ha mixed ammonium salts | 23.5% N |
|    | 1916–67 | 689 kg/ha sulphate of ammonia[a] | 21% N |
|    | 1968–85 | 686 kg/ha 'Nitro Chalk'[c] | 21% N |
|    | 1986–89 | 417 kg/ha 'Nitram' | 34.5% N |
| N4 | 1968–85 | 914 kg/ha 'Nitro Chalk'[c] | 21% N |
|    | 1986–89 | 556 kg/ha 'Nitram' | 34.5% N |
| P  | 1852–88 | 392 kg/ha single-superphosphate[b] | 18% $P_2O_5$ |
|    | 1889–97 | 407 kg/ha single-superphosphate | 18% $P_2O_5$ |
|    | 1989–1902 | 448 kg/ha basic slag | – |
|    | 1903–73 | 407 kg/ha single-superphosphate | 18% $P_2O_5$ |
|    | 1974 | 161 kg/ha triple-superphosphate | 47% $P_2O_5$ |
|    | 1975–86 | 407 kg/ha single-superphosphate | 18% $P_2O_5$ |
|    | 1987–89 | 161 kg/ha triple-superphosphate | 47% $P_2O_5$ |
| K  | 1852–58 | 336 kg/ha sulphate of potash | 49% $K_2O$ |
|    | 1859–1989 | 225 kg/ha sulphate of potash | 49% $K_2O$ |
| Na | 1852–1973 | 112 kg/ha sulphate of soda | 14% Na |
| Mg | 1852–1973 | 112 kg/ha sulphate of magnesia | 10% Mg |
|    | 1974–89 | 208 kg/ha kieserite applied every third year | 16.8% Mg |
| FYM | 1852–1989 | 35 tonne/ha FYM | – |

Notes for Tables 9.2–9.4.
[a] Mixed ammonium salts consisted of ammonium sulphate and ammonium chloride in equal proportions.
[b] Until 1888, single-superphosphate was made on the farm by mixing bone dust with sulphuric acid.
[c] The concentration of N in Nitro Chalk increased from 21 to 27.5% and the amount supplied was decreased to maintain a constant rate of N application.
[d] On Broadbalk, no fertilizers or FYM were applied in fallow years before 1968. Since 1968, fertilizers other than N and FYM have been applied to fallow sections.

during this period: in 1956 section I was split into sections IA and IB. Section IB continued in the five-year rotation while section IA grew continuous wheat with use of weed-killers as required. In 1963 section V was divided in the same way, section VA continued in the rotation while VB dropped out. In 1968 the fallowing scheme was stopped, sections II, III and IV split into two and the 10 sections were established as they are known today (section 0 being the old section IA, section 9 the old section VB). Since then the different sections have been used to investigate various cropping regimes.

In our study sections 1, 2, 4 and 7 were examined for each plot chosen. On section 1 wheat was grown continuously except during fallow years between 1935 and 1968. During this period all sections of the plot were included in the analysis and the five-year cycle was treated as a regular rotation, taking average inputs and yields over the whole cycle (details can be found in Section 9.4.2). Sections 2, 4 and 7 took part in a wheat–potatoes–beans rotation from 1968–78 and were considered for these years only.

## 9.3 DATA COLLECTION: SOURCES, PROBLEMS AND METHODS

In order to calculate the TFP/TSFP economic indexes, it is necessary to assemble and quantify the input and output factors.

The factors considered as *inputs* to crop production were: *the cost of buying or renting of land; the cost of employing labour, machinery* and, in earlier years, *horses* to cultivate the land; *the cost of seed* where necessary; *the cost of any fertilizer or lime* used to improve the soil and the cost of any *herbicides, pesticides or fungicides* used to protect the crop. Inclusion of land costs as an input is arguable, since by definition it is a fixed input and in contrast to the other inputs does not have a direct marginal effect on the outputs of the agricultural system. Nevertheless, land, by providing a flow of services, has a positive shadow price and this is approximated by the rent paid for the use of these services. Moreover, particularly in recent years, the value of land may influence the farmer's decision on whether to farm the land or utilise it in some other way (diversification) and so we believe that this factor must be included in the analysis.

From a purely economic standpoint, the only *output* from the system is the *crop yield*. However, in order to calculate the TSFP, other outputs from the system—the externalities such as pesticide residue, damage to wildlife and nitrate leaching—must also be considered. Since complex issues are involved in costing these factors and little quantitative data exists, they will be considered separately later.

The indexes we considered are based on quantities weighted by cost shares of the respective input or output factor. To determine these cost shares, the quantity used and the price per unit is needed for each of the factors. All data were appropriate to the harvest year and all quantities were standardised to refer to one hectare.

Since the Rothamsted experiments date back to the middle of the last century, accurate cost data for the various factors have in many cases been difficult to find. The lack of availability of these data has forced us to omit the earliest years from the experiments. For this reason we have analysed the Park Grass experiment from 1875 onwards, and Broadbalk from 1871 onwards. Although as much cost data as possible were collected, some relatively large gaps remain. Data collection consisted of two parts: obtaining any exact data available, and estimation of the missing values in order to render the data suitable for analysis.

### 9.3.1 Land

It was decided to treat the Rothamsted land as if it was rented, since the cost of renting land should reflect land prices. Yearly average rents in England and Wales (£/ha) from 1871–1990 taken from Lloyd (1992) were used.

The missing rent price for 1991 was estimated by calculating the increase in rent prices over the three previous years, and using the average as an inflation factor for 1990–91. Although this method would not be satisfactory for a series of rent prices, it seems reasonable for a single value.

### 9.3.2 Labour

The average wages and number of hours worked per week for hired manual labour from 1903–92 were found in the following publications: Broomhall (1903, weekly hours and earnings), Marks (1989) (1914, 1917–66, 1969, 1971, 1974, 1976 weekly earnings and 1914–46 weekly hours), Ministry of Agriculture, Fisheries and Food (MAFF) (1968) (1949–69 weekly hours) and Nix (1967–92 weekly hours and earnings, where the average weekly earnings were estimates based on the previous year). Where the number of hours worked to achieve the weekly wage was not found, this was estimated by linear interpolation from the available data. A series of labour costs per hour could be generated from 1903–92, missing only 1904–13, and 1915–16. A comparison of rent and labour costs over time showed very similar patterns. It therefore seemed sensible to predict missing labour costs from a linear regression of labour costs on rent prices.

As well as the cost of labour, the hours of labour required to produce the crops considered, i.e. hay (1875–1991), winter wheat grain and straw (1871–1988), beans (1968–78) and potatoes (1968–78), were required, Nix (1967–92) gives labour requirements for each of these crops. Various editions of Watson and More (1924–49) cite unchanging labour requirements for cereals and hay over that period. These constant figures were therefore used for all years 1924–49.

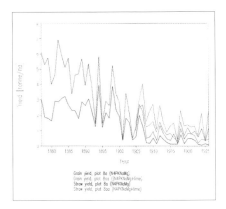

**Figure 9.1** Yields from the Woburn continuous wheat experiment, 1877–1926.

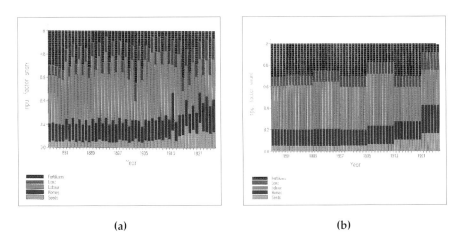

(a)  (b)

**Figure 9.3** (a) Factor cost shares based on prices from each year for Woburn plot 8a (*N4PKNaMg*); (b) factor cost shares based on nine-year average prices for Woburn plot 8a (*N4PKNaMg*).

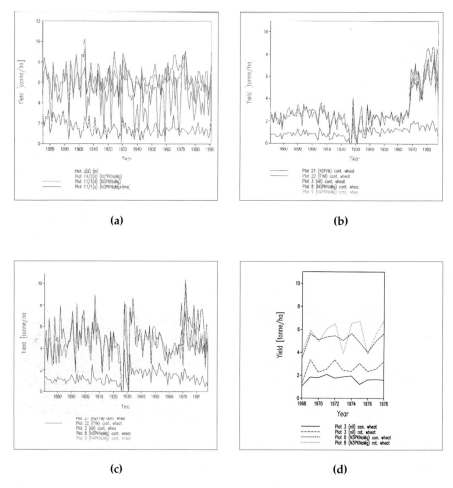

**Figure 9.4** (a) Hay yields from Park Grass experiment, 1875 – 1991; (b) wheat grain yields from Broadbalk, 1871 – 1988; (c) wheat straw yields from Broadbalk, 1871 – 1988; (d) wheat grain yields from rotational or continuous wheat on Broadbalk, 1968 – 1978.

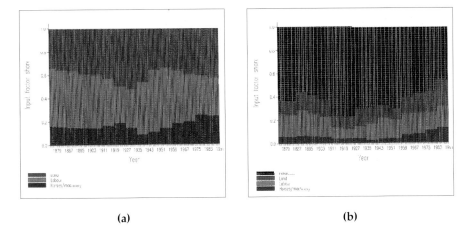

**Figure 9.7** (a) Factor cost shares for Park Grass, plot 3d (nil fertiliser); (b) for Park Grass, plot 14/2d (*N2*PKNaMg*).

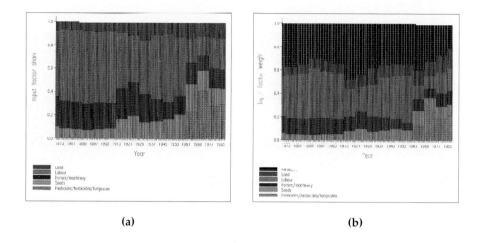

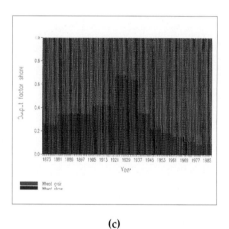

**Figure 9.11** (a) Factor cost shares for Broadbalk, plot 3 (nil fertiliser); (b) for Broadbalk, plot 8 (*N3PKNaMg*); (c) factor revenue shares for wheat grain and straw from Broadbalk, plot 8 (*N3PKNaMg*).

This left missing data for labour requirements for production of wheat grain (1871–1923 and 1950–60), the extra production cost associated with harvesting wheat straw (1871–1966), and for the production of hay (1875–1923 and 1950–60). For the Broadbalk plots left to fallow, labour requirements for wheat grain and straw were set to zero in the respective years.

Labour requirements for the production of hay and wheat before 1924 were taken as the 1924 values, since changes in technology were relatively small over the last part of the nineteenth century. However, labour requirements for 1950–60 could not be expected to have stayed constant, and so were estimated by linear interpolation from values immediately before 1950 and after 1960, i.e. assuming steady change in technological improvements over this time period. Due to the lack of data on labour requirements for wheat straw, neither of these methods could be used, but because of the loose relationship between wheat grain and straw production, missing values were predicted from a linear regression of the present values on wheat grain production figures. It was also known (information from Rothamsted farm) that the additional labour required for harvesting straw took about 30% of the total labour required for growing wheat between 1880 and 1930.

For fallow years, the labour required to produce wheat grain and wheat straw has been set to zero since no wheat was grown in those years. However, some labour would still be required for general management and for the summer cultivations used to suppress weeds. Therefore for fallow years, 70% of the labour requirement needed to produce wheat grain has been used to take account of the necessary work.

Since these labour requirements assume a 'standard' set of operations, they must be manipulated for any change in circumstances. It was assumed that the 'standard' labour requirements included components for fertilizer application, lime application and pesticide application (after 1960). Adjustments therefore had to be made when farmyard manure (FYM) was applied, fertilizers or lime were not applied or pesticides were not used (after 1960). Since the absolute amount of time required for any of these operations is related to the available equipment, a proportionate cost was added to or removed from the labour requirements. These proportions were established using present day estimates: for the proportion of time used applying fertilizers and lime out of the total labour requirement, Nix (1992) was used. An estimate of 2 hours was obtained (B. D. Prew, Rothamsted, personal communication) for the application of FYM to one hectare—this was used to generate a proportionate cost of applying FYM. The labour needed for spraying pesticides was included in Nix's cost of standard applications and was therefore taken account of through the weed-killer factor. Once the labour requirement data series was established an allowance was made for the reduced labour for the fallow years on some of the Broadbalk plots and the requirements were adjusted for any non-use of fertilizers, non-use of farmyard manure or non-use of lime as described.

### 9.3.3 Horse and machinery

A survey of general farm management textbooks (see for example Watson and More, 1924–49) led us to the follow basic rules for accommodating the change from horses to tractors. From the middle of the nineteenth century until 1930, farming methods were based on horse power alone. Between 1930 and 1959 tractors and horses shared the workload, with the proportion of work done by tractor increasing until it took over completely from horses around 1960. Watson and More presented figures showing changes in the division of labour between horses and tractors over these years: in 1924 no tractors were used; in 1933 the maximum fraction of the work that could profitable be performed by a tractor was 1/3 (and one tractor could do the work of five horses); this rose to 2/3 in 1941 (when a tractor could replace seven horses) and to 3/4 in 1949. Therefore, the cost of using horses was needed from 1871 to 1959 and the cost of using tractors was needed from 1930 onwards. Of course, tractors are not the only machinery used for crop production, but they represent a major component of the mechanical work performed and can be used as a standard for machinery costs.

An estimated cost of horse labour per hour was found in Broomhall and in Watson and More. The cost per hour for running a tractor was found in Watson and More and Nix (1967–91) using values for a 45–50 kW machine. This leaves large gaps in the data for horse costs (1871–1902 and 1904–32) and tractor costs (1950–60).

The number of hours of labour from horses or a tractor needed to produce the various crops was also required. Watson and More present the number of horse-days (assuming an eight-hour working day) to produce cereals and hay. Using their guidance (described above) for the conversion from horse to tractor power provides both horse and tractor requirements for wheat grain and hay from 1924–49. From 1967–92, tractor requirements for all the crops we considered were available from Nix's 'Farm Management Book'. This leaves gaps from 1871–1923 for wheat and hay, where it does not seem unreasonable to assume constant horse requirements, and between 1950 and 1960. The marginal requirements for straw production (on top of grain production) from 1871–1966 had to be estimated. For the Broadbalk plots in fallow, tractor and horse requirements for wheat grain and straw were set to zero in the respective years. Missing prices for the cost of running a horse or tractor per hour were estimated by a linear regression on rent prices, as a default inflation factor.

Before 1924, the required amount of horse labour to produce a crop is, like human labour, assumed constant due to relatively small technological change over the period. Later gaps in horse or tractor requirements were estimated by linear interpolation between known values, again assuming a steady improvement in technology. Requirements for wheat grain production were again used to estimate the marginal costs for wheat straw production, given the overlap between the two activities. Applying the same argument used for human labour

## Data collection: sources, problems and methods 183

in fallow years, horse and tractor requirements for wheat straw in fallow years were set to zero but field management work was assumed to take 70% of the work necessary to produce grain.

### 9.3.4 Fertilizer, FYM and lime

On the long-term experiments analysed, the fertilizer treatments consisted of different combinations of single nutrient fertilizers supplying nitrogen (N), phosphorus (P) and potassium (K). Nitrogen was applied either as sulphate of ammonia, nitrate of soda, ammonium nitrate as 'Nitro Chalk' or, in recent years, ammonium nitrate as 'Nitram'. Phosphorus has been applied as single-superphosphate until very recently when triple-superphosphate was used. Until 1888 the single- superphosphate was made from bone ash and sulphuric acid. In a few years around 1900, basic slag was used instead of single-superphosphate. Potassium has been applied always as sulphate of potash. For full details of the fertilizer treatments on the plots considered here see Tables 9.2–9.4. As well as the major nutrients, some plots also had small amounts of sodium sulphate and magnesium sulphate applied. Since no prices for these components could be found for any of the years considered, and because they were of minor importance, they were excluded from the analysis.

The quantity of fertilizer applied was usually adjusted over the period of the experiments so that the amount of element applied per unit area remained constant over time. Consequently changes in both the amount and cost of fertilizer occured. The amount of FYM has always been the same—35 t/ha fresh material—and the total nitrogen content has remained remarkably constant.

For 1856–1910, a number of fertilizer prices were retrieved from publications such as 'The Gardener's Chronicle' (Anon. 1856–70), 'The Agricultural Gazette' (Ann. 1871–1901) and from the 'Gentlemen's Estate Book' (Broomhall, 1903, 1907). The percentage of active constituents in the fertilizer was sometimes also given. For 1911–21, yearly average fertilizer prices appeared in the MAFF 'Agricultural Statistics'. For 1922–91 August prices (or July prices if necessary) were collected from MAFF's 'Agricultural Statistics'. For 1940 to 1947, average annual data were obtained from the Central Statistical Office (CSO) publication 'Annual Abstract of Statistics'. Some occasional values were extracted from Watson and More (1924–49, intermittent). Finally, Nix (1967–92) gave yearly fertilizer prices. This gave an intermittent set of fertilizer prices with no large gaps.

No lime prices were available, except as five-yearly prices from 1955 onwards. A similar situation arose with respect to the cost of farmyard manure (FYM). Since farms produced their own FYM, no record of prices could be found.

Fertilizer prices (for fixed percentages of the active constituents) were calculated using either known or estimated nutrient percentages of products to standardise prices. Gaps within the fertilizer price series were estimated from the values known. For single-superphosphate, where a reasonable number of values were available, prices were estimated by linear interpolation between known values. This procedure was also used for nitrate of soda before 1888. Later missing values for nitrate of soda were predicted from linear regression on single-superphosphate prices. The four absent values for triple-superphosphate were estimated by linear interpolation between known values. This was also the procedure for estimating prices of sulphate of potash after 1940. Before 1940, no prices were available for sulphate of potash. The remaining values were constructed using a linear relationship with nitrate of soda (which had also been partially estimated). Further application of linear interpolation completed the basic slag, bone ash, sulphate of ammonia and 'Nitro Chalk' price series. The very low price for sulphuric acid was held constant over the few years it was used.

Although these methods led to a set of interrelated fertilizer prices, this does not seem unreasonable and produced prices which appear plausible. The alternative approach of using only years when all data were available left very few years to be analysed and was discarded as impractical.

Where the price of lime was available, it was used to establish the cost of lime as a proportion of total labour costs and this formula was employed as a token price of lime. In order to include a token cost for FYM, a cost of production equal to the cost of applying FYM to the field was imposed. Again, this was not entirely satisfactory, but allowed us to attribute some component of the overall cost to lime and FYM.

### 9.3.5 Seed

Seed cost was required for winter wheat on Broadbalk and Woburn, but not for Park Grass, which had been in permanent grass for at least 200 years in 1856. Occasional seed prices were found in 'The Agricultural Gazette' and Broomhall's 'Gentlemen's Estate Book'. From 1911–32, September prices (chosen because they were consistently available) for wheat seeds were taken from the MAFF 'Agricultural Statistics'. Nix (1967–92) supplied seed price per hectare for wheat (1968–88), and potatoes and beans (1968–78). These prices were converted back to price per tonne from the seed rates used in the experiment.

Once again linear interpolation provided the missing data for seed potatoes and bean seed. Wheat seed prices were calculated as a proportion of wheat grain prices. Unfortunately this relationship was not constant because wheat seed became increasingly more expensive relative to wheat grain after the 1960s. We therefore used two regression lines—one before and one after 1960—to get a full set of wheat seed prices. Our estimated price increase for seeds over grain was 1.17 (with a standard error (se) of 0.08) before 1960 and 2.00 (se 0.06) after 1960.

### 9.3.6 Herbicide, fungicide and pesticide

Like many farms, Rothamsted started to use chemical means to control pests and diseases in the 1960s. Therefore an input cost for pesticides, fungicides and herbicides had to be included in the analysis for Broadbalk, the only experiment considered on which they were applied. The exact quantities of each compound used on the experiments can be found in the Rothamsted Experimental Station 'Yields of the Field Experiments' (1855–1991). Unfortunately, the only available source of pesticide costs (Nix, 1967–92) quoted cost per standard application without further explanation. The following method was used as an attempt to construct realistic prices. Where compounds were applied individually they were treated as one 'standard application'; where two or more compounds were applied together, the combination was treated as one 'standard application' with its price calculated as the average price of the constituent compounds. Since smaller quantities of chemicals were used for multiple than for single chemical applications this approach seemed justified. However, Nix did not provide data for every compound used or for every year, so some missing values remained.

Since a 'plant protection product' inflation index (based on 1964 prices) was available from the CSO 'Monthly Digest of Statistics' (1964), these inflation factors were used to estimate missing prices when any prices for the compound were given. Where no 1964 price was available for a product, this was estimated by converting all available values back to 1964 via the index, then taking the average. The compounds for which no data at all were available through Nix were left out of the analysis.

### 9.3.7 Yield

From a solely economic standpoint, crop yield represents the only output factor. For the experimental plots we have considered, outputs are yields of wheat grain and straw (1871–1988), maincrop potatoes and beans (1968–78) from Broadbalk, wheat grain and straw (1877–1926) from Woburn continuous wheat, and hay (1875–1991) from Park Grass.

Yields of the experimental plots chosen were available from Rothamsted records; see Rothamsted Experimental Station (1855–1991). Occasional values were absent due to pest damage or errors in experimental procedure. These included some wheat yields from Woburn plot 8a and some bean yields from the Broadbalk rotational plots 21, 22 and 3. Furthermore, 1987 straw yields from Broadbalk continuous wheat on plots 3 and 8 were not given in these reports and so the figures were recalculated from the fresh weights recorded in Rothamsted files.

On Park Grass a new harvesting method was introduced in 1960. Park Grass is cut twice, once for hay in June/July and then in October/November. Until 1959, this first cut was made into hay on the plot, and the yield recorded as hay which could be converted to dry matter. From 1960 onwards, yields of the first cut have been estimated from either two or four cuts made by a forage harvester, the remainder being cut and made into hay on the plot in order to maintain continuity of husbandry. The fresh weight is determined and also the dry matter in the fresh sample; yields are recorded as dry matter. Because there have been no dry matter losses associated with hay making since 1960 this leads to higher estimates of yield. An analysis of Park Grass yields (J. M. Potts, Rothamsted, personal communication) estimated the factor by which estimated yields were inflated using the new harvesting methods. This factor was used to standardise, yields to the pre-1960 method.

Prices for all relevant crops were obtained from 'The Agricultural Gazette' (1874–1901) and from the MAFF 'Agricultural Statistics' (1866–1988). The Agricultural Gazette prices represent market prices from different times of the year. The MAFF prices are January prices (except wheat grain in 1871–91). This set of crop prices is nearly complete, with only a few missing values; hay prices were missing in several years before 1900 and in several recent years; straw prices were absent in nine years; wheat grain was missing in one year; but the bean and potato price series were complete.

The few missing yield values were predicted using linear regression: the absent grain and straw yields from Woburn plot 8a were predicted from a linear relationship estimated between the yields from plot 8a and plot 8aa–whose treatment differed from that of plot 8a in the application of lime; the missing bean yields from Broadbalk rotational plots 21, 22 and 3 were predicted from a multiple regression on bean yields from plots 8 and 9. It was expected that similar plots would produce roughly similar yields over time, and the year effect on one plot would be expected to be equivalent to the year effect on any other plot in the same experiment.

The hay price series was completed using linear interpolation. This was considered acceptable since it involved only one or two values missing in the series before 1900. The four missing values at the end of the series were predicted by the average of the three previous years' prices. This was dictated by the pattern of the data, which seemed to indicate a decrease in hay prices rather than the increase due to inflation one might expect. The single absent wheat price was predicted by linear interpolation. Finally, the missing straw prices were established assuming a linear relationship with grain prices. Although straw as a crop has decreased in relative value, this method still seemed to produce reasonable prices.

## 9.4 TFP INDEXES EXCLUDING EXTERNALITIES

Assessment of the sustainability of an agricultural system requires mesurement of productivity at any single timepoint as well as measurement of trends in productivity over time. The measurement of the productivity of a system involves a comparison of the outputs produced from the agricultural system in relation to the inputs utilized. Total factor productivity (TFP) is the ratio of an index of aggregate output to an index of aggregate input. By constructing an index of TFP over time, it is possible to assess the long-term productive performance of the system.

The investigation of the TFP indexes for the experimental plots was carried out in two stages: first an assessment of the plots excluding externalities, and secondly including externalities. This section covers the work on TFP indexes excluding externalities.

### 9.4.1 The choice of a TFP index

The performance of a number of different economic indexes was compared over the three experiments. We considered three arithmetic and three geometric indexes using a variety of methods to construct price weights. For all indexes, the final timepoint was defined to be 1.0. We examined chain-linking and also direct comparison with the end-point using a fixed price set.

*Arithmetic indexes*

Two arithmetic indexes were considered which used constant prices from either the end (Index $A1$) or the beginning (Index $A2$) of the series of data and related values directly to the end of the series.

At time $t$, the value of index $A1$ for the output series only (running from year 0 to year $T$) is defined by

$$A1_t = \sum R_{jTT}(Q_{jt}/Q_{jT}); \qquad R_{jTT} = P_{jT}Q_{jT}/(\sum P_{kT}Q_{kT})$$

where

$Q_{jt}$ is the output quantity for factor $j$ at time $t$
$P_{jt}$ is the price of factor $j$ at time $t$
$R_{jtu}$ is the $j$th factor revenue share based on prices at time $t$ and quantities at time $u$.

The input index is constructed similarly using cost shares and quantities for input factors, and the overall TFP index is a ratio of the output index to the input index. This index is similar to the Laspeyres index using current prices, except that the base period is the end of the series rather than the beginning.

The second index, $A2$, is similar to the Paasche index in that it uses price weights from the beginning of the series, but again it relates directly to the end of the series, time $T$. The output index then becomes

$$A2_t = \sum R_{j0T}(Q_{jt}/Q_{jT}) \qquad \text{where} \qquad R_{j0T} = P_{j0}Q_{jT}/(\sum P_{k0}Q_{kT}).$$

The only difference between indexes *A1* and *A2* is in the calculation of factor cost shares. These two indexes are theoretically unsatisfactory since they use constant prices to relate directly to the end-point. Prices are used to define the weights $R_{jtu}$ on the assumption that they reflect the 'true' value of each component relative to other components. Over the long period we are considering (up to 118 years) we know that relative values have changed and we need to take account of this. Unlike direct comparison with the end-point, chain-linking relates year $t$ to year $T$ via all intervening years using updated prices, and is generally acknowledged to give better results. We therefore constructed a third index taking these points into account.

The third arithmetic index (*A3*) compromised by changing the price weights every $K$ years, then chain-linking through successive time periods to relate to the end-point. This was done as follows: each series of measurements was divided into $L$ overlapping subseries each with $K$ values, and each subseries had one point in common with the subseries before and one point in common with the following subseries. Let $(l, k)$ be the $k$th time within series $l$, then $0 = (1, 1)$; $T = (L, K)$ and $(l, K) = (l + 1, 1)$. For each subseries, average prices are used, e.g. for subseries $l$, an average price $\tilde{p}_j^l = \sum_k (p_{jk}^l)/K$ over the period is calculated for each factor. Within subseries $l$, the TFP index at time $(l, k)$ is then calculated using the constant prices $\tilde{p}_j^l$ to relate time $(l, k)$ to time $(l, K)$. Let $A3_{l,k;l,K}$ be the value of the index (considering outputs only) at time $(l, k)$ related to time $(l, K)$:

$$A3_{l,k;l,K} = \sum R_{j,K}^l (Q_{jk}^l / Q_{jK}^l)$$

where

$Q_{jk}^l$ is the output quantity for factor $j$ at time $lk$

and

$$R_{j,K}^l = \tilde{p}_j^l Q_{jK}^l / (\sum \tilde{p}_i^l Q_{iK}^l).$$

So $R_{j,K}^l$ is the $j$th factor cost share within series $l$ using average prices within the subseries and quantities from the last point of the subseries. The index then relates timepoints to the final value T by chain-linking:

$$A3_{lk} = A3_{l,k;l,K} \cdot A3_{(l+1)1;(l+1)K} \cdots A3_{L,1;LK}$$
$$= A3_{l,k;l,K} \cdot A3_{lk;(l+1)K} \cdots A3_{(L-1)K;LK}$$

since the overlapping series mean that $(l + 1, 1) = (l, K)$.

If the subseries are of length 2, then this reduces to normal chain-linking for an arithmetic index. Since our weights, or factor shares, changed gradually over time an update of prices every nine years seemed sufficient to reflect changes in relative weights and hence subseries of length nine were used.

## Geometric indexes

A geometric index (for output quantities) is of the form

$$\ln\{G_t\} = \sum R_{juT} \ln(Q_{jt}/Q_{jT}).$$

Again, we considered three possible indexes, calculated analogously to the three arithmetic indexes: with constant prices fixed at the beginning and end of the series referred directly to the end-point; and a chain-linking geometric index using constant averaged prices within nine-year subseries and updating prices between subseries.

## Comparison of indexes

To illustrate the different indexes, Figure 9.1 shows the yield, and Figures 9.2a–b show the six indexes considered for Woburn continuous wheat plot 8a. To give further insight, Figure 9.3a presents input factor cost shares at current prices and Figure 9.3b shows the shares used to construct the $A3$ and $G3$ indexes. The whole set of indexes ($A1-3, G1-3$) was calculated for other plots, with similar results. For a given plot there was little difference between the three arithmetic indexes or between the three geometric indexes. Although the arithmetic and geometric indexes were clearly on different scales they showed the same basic pattern, closely related to yield.

We believe that the similarly between the various indexes occurred because of the reasonably stable nature of the factor shares (see Figure 9.3a for input factor cost shares at contemporary prices; on this plot changes in the fertilizer shares are caused by the use of sulphate of ammonia as nitrogen fertilizer every second year). Differences between indexes are based mainly on differences in factor shares. Clearly, if factor cost shares are reasonably steady, then the indexes will be similar. However, although the shares did not change drastically, there was a steady trend over the time period in the changing factor shares, and, to take account of this, updating prices and chain-linking seemed the more sensible option. Since there was no obvious difference in interpretation between the arithmetic ($A3$) and geometric ($G3$) indexes over various plots, the arithmetic index $A3$ was chosen as the index to be used for further comparisons.

It should be noted that, although the arithmetic chain-linked index is adequate in our situation, these results would not necessarily carry over to other situations. Our indexes are calculated on long-term experiments where most of the inputs have, almost by definition, been held constant over the entire time period. In other situations, treatment of the site might well be less consistent, possibly leading to quite variable factor shares. An index such as Tornqvist–Theil, using average factor cost shares and the more theoretically flexible approach of the geometric function, might then give different and perhaps more appropriate results. Here, with quite stable factor shares, the use of average factor shares would make little difference.

## 190  Sustainability—the Rothamsted experience

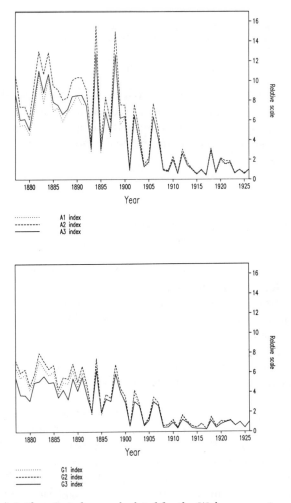

**Figure 9.2** (a) Arithmetic indexes calculated for the Woburn experiment, (b) Geometric indexes calculated for the Woburn experiment

It was also noted above that our indexes tend to follow yield patterns quite closely (further yield plots are given in Figures 9.4a–d). We believe this also arises from the consistent treatment of the experiments over time, which would lead to a reasonably stable input index. Since the output index consists of yields alone, which fluctuate substantially between years, the output index will closely follow yields and swamp the smaller changes in the input index, giving a TFP index which also closely follows yield. Again, this might well change under a more variable management regime.

### 9.4.2 Performance of experimental plots over time

Our first concern was to use the TFP indexes to assess the economic sustainability of each of our experimental plots and compare the results with our prior expectations. We chose the arithmetic index $A3$ as our measure of performance over time, *not yet including externalities*.

However, we had serious reservations about using the index $A3$ alone as a measure of sustainability. An intuitive definition of sustainability would suggest two criteria:

(i) that the output from a system can be maintained over time;
(ii) that the agricultural system produces output on some scale (e.g. money or energy) at least comparable with, and preferably in excess of, the inputs to the system.

The index $A3$ provides a good measure of the first criterion, of whether the trend of the output/input ratio of the system is increasing or decreasing. But it provides no information on the absolute value of the output/input ratio, i.e. whether the value of resources put into the system outweighs the value of the crop they produce. In an attempt to measure the balance of inputs to and outputs from an agricultural system, we have used the output/input ratio (excluding rents), i.e. the output/input ratio including all factors directly relevant to the yield production from the land. Unlike index $A3$, the output/input ratio is based on contemporary prices and is therefore not easily interpretable in terms of trend over time, only as an absolute value at individual time points. The inclusion of a measure to check the second criterion above allows easy comparison between systems with similar trend but very different output/input ratios.

Our working definition is then: an agricultural system is regarded as economically sustainable if it is profitable (output/input ratio above 1) and does not show a downward trend over time ($A3$ index).

A summary of our results in applying these measures can be seen in Table 9.5, and these results are discussed in detail below.

#### *Park Grass*

The $A3$ index and output/input ratio described above were calculated for the four plots selected from the Park Grass experiment. The results largely matched our expectations (see Section 9.2.1) and are shown in Figures 9.5 and 9.6. The arithmetic index showed no clear trend over time for any of the plots. The output/input ratios suggest that on plots 14/2(d) and 11/1(a), revenue clearly exceeds input costs, whilst on plots 11/1(d) and 3(d), the ratio is very variable, but on average output costs appear approximately equal to input costs. This seems to reflect what has actually happened on Park Grass. Yields for plot 3(d) are very low, but there are few inputs. Since these yields are stable, it seems reasonable that the $A3$ index shows no trend. Our prior conception was that plot

**192**  *Sustainability—the Rothamsted experience*

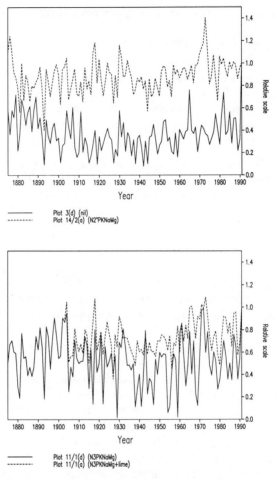

**Figure 9.5**  *A3* index calculated for various plots on Park Grass (excluding externalities)

11/1(d) might not be considered sustainable, since application of fertilizer and no lime has reduced the soil pH to 3.5. Yields have however been maintained on this plot (albeit at a lower level than on 11/1(a)) since the plot is now dominated by one acid-tolerant species. This also means that yields are more variable than on other plots since the single species is more sensitive to adverse conditions than a mixture of species. Despite this, the plot appears just to achieve economic sustainability although this kind of instability would not be a desirable feature for a sustainable system. However, inclusion of externalities would surely need to include some appreciation of biodiversity and soil health, which might change these conclusions.

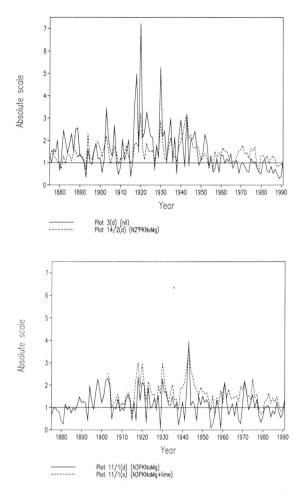

**Figure 9.6** Output/input ratio (productivity) for various plots on Park Grass (excluding externalities)

The input factor cost weights used to construct the *A3* index illustrate the cost weights for plot 3(d) where no fertilizers were applied (see Figure 9.7a) in contrast to plot 14/2(d) where fertilizers were used (see Figure 9.7b).

*Woburn*

The assessment of economic sustainability for the Woburn plots gave exactly the results expected (see Figure 9.2a and Figure 9.8). Both the index and ratio measures show these plots are markedly unsustainable, with the arithmetic indexes showing a steep downward trend and the cost ratios indicating unprofitabe systems (outputs by far outweighed by inputs). This illustrates what

**Table 9.5** Assessment for economic sustainability using index $A3$ and the output/input ratio

| Plot | Arithmetic index $A3$ | Output/input ratio | economically sustainable? |
|---|---|---|---|
| Part Grass | | | |
| Plot 3(d) | no trend | around 1 | yes |
| Plot 11/1(a) | no trend | higher than 1 | yes |
| Plot 11/1(d) | no trend | around 1 | yes |
| Plot 14/2(d) | no trend | higher than 1 | yes |
| Woburn Continuous Wheat | | | |
| Plot 8a | downward trend | lower than 1 | no |
| Plot 8aa | downward trend | lower than 1 | no |
| Broadbalk Continuous and Rotational Wheat | | | |
| Plot 11 continuous wheat | no trend | higher than 1 | yes |
| Plot 22 continuous wheat | upward trend | higher than 1 | yes |
| Plot 3 continuous wheat | upward trend | around/below 1 | no |
| Plot 8 continuous wheat | upward trend | higher than 1 | yes |
| Plot 9 continuous wheat | no trend | higher than 1 | yes |
| Plot 21 rotational wheat | no trend | higher than 1 | yes |
| Plot 22 rotational wheat | no trend | higher than 1 | yes |
| Plot 3 rotational wheat | no trend | higher than 1 | yes |
| Plot 8 rotational wheat | no trend | higher than 1 | yes |
| Plot 9 rotational wheat | no trend | higher than 1 | yes |

eventually led to the end of this experiment: continuous wheat was grown on the experiment year after year, and the use of ammonium sulphate as a nitrogen fertilizer without lime led to s sharp increase in the soil acidity, which in turn led to decreasing yields. An attempt was made in 1905 to reverse the process by liming the plot 8aa to maintain a higher pH. However, as the $A3$ index and the output/input ratio for plot 8aa show, this attempt failed.

The input factor cost weights used to construct the $A3$ index (see Figure 9.3b) show increasing cost weights for seeds. The changes in cost weights for fertilizers depend on whether sulphate of ammonia was applied at the end-points of the subseries. This varies since the fertilizer was applied every second year.

### *Broadbalk*

In order to take account of the regular fallow scheme on Broadbalk from 1935–67 (see Section 9.2 for details) we based our $A3$ indexes and input/

*TFP indexes excluding externalities* 195

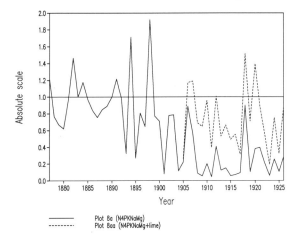

**Figure 9.8** Output/input ratio (productivity) for Woburn Continuous Wheat experiment (excluding externalities)

output ratios for these years on sections I–V rather than on section I alone. This affects plots 23, 3 and 8 only. These plots were treated as one hectare fields in which a rotation–four years wheat then one year fallow–took place on five equal field sections. The yield for each year was calculated as an average over the five sections, including the zero yield from the fallow section. Similarly, the input cost was calculated as an average of the input costs from all five sections. This method was more appropriate than considering a single section over the whole time period since the effects of the fallowing (reduction of weeds and supply of nitrogen) are spread over the whole cycle, which gives a realistic

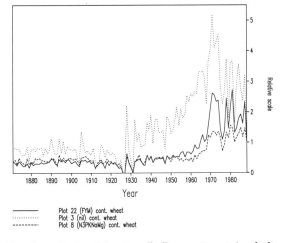

**Figure 9.9** *A3* index calculated for Broadbalk experiment (excluding externalities)

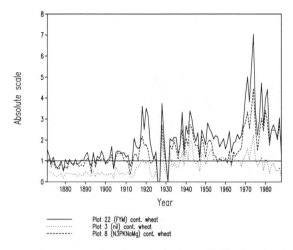

**Figure 9.10** Output/input ratio (productivity) of Broadbalk plots (excluding externalities).

description of a rotational farming system where a proportion of the land would be fallow at any one time.

Using the $A3$ index and output/input measure, all the Broadbalk plots considered (apart from plot 3, continuous wheat) were judged to be economically sustainable. For plots 22, 3 and 8, which were included for the full length of the experiment, the $A3$ indexes are constant until the 1930s, when a clear upwards trend begins; see Figure 9.9. In terms of profitability, the output/input ratios suggest that for all treatments except plot 3 continuous wheat (no fertilizer treatment), the outputs easily exceed the inputs; see Figure 9.10.

The upward trend shown in the $A3$ indexes reflects the increasing grain yield from the 1930s onwards. There are several possible explanations for this trend. In 1926/1927, periodic fallowing was introduced on Broadbalk to control weeds. After 1964, herbicides, fungicides and insecticides were routinely used to control plant and animal pests. New varieties of wheat were also brought into the experiment.

Finally, the cost shares used to construct the $A3$ index illustrate again how input weights change when fertilizers are applied (see Figure 9.11b) in contrast to an untreated plot (see Figure 9.11a). Assessment of output factor cost weights also shows a steady increase in wheat grain cost shares from the 1920s/30s onwards. (See for example the output cost shares for plot 8, continuous wheat in Figure 9.11c.)

The rotational wheat plots (1968–78) and other plots where treatments started after 1968 were not assessed for trend over time since the series were so short.

### 9.4.3 Comparisons between plots

In order to directly compare different treatment regimes, output/input ratios were related across different plots. A measure for comparing outputs from plot A with plot B at time $t, t = 1, \ldots, T$ is given by

$$\sum R_{jtB}(Q_{jtA}/Q_{jtB}) \quad \text{where} \quad R_{jtB} = P_{jt}Q_{jtB}/(\sum P_{kt}Q_{ktB})$$

where

$Q_{jtA}, Q_{jtB}$ is the output quantity for factor $j$ at time $t$ on plots A and B respectively,
$P_{jt}$ is the price of factor $j$ at time $t$,
$R_{jtA}, R_{jtB}$ is the $j$th factor revenue share based on prices and quantities at time $t$ from plots A and B respectively.

A measure to compare inputs to different plots can be defined analogously. The ratio of the output and input measures gives a comparison of productivity between the two plots. This is in fact just the ratio of output/input ratios for plots A and B. A value smaller than one shows that plot B is more profitable than plot A. Since prices are neither fixed nor updated, no reliable statement can be made about trends over time using this measure.

An index for comparing pairs of plots over time could easily be constructed using either fixed prices and direct comparison to the end-point (e.g. indexes *A1* and *A2* in Section 9.4.1) or the use of updated weights combined with chain-linking (as for index *A3*). Another approach would be direct comparison of a set of *A3* indexes for different plots, standardizing all indexes by the final value of one plot, the base plot. Only the base plot would then be defined as 1.0 at the endpoint, and useful comparisons could be made.

*Park Grass*

The Park Grass yield plots (see Figure 9.4a) show that although all plots seem to follow the same yield pattern, hay yields from plot 3(d) are markedly lower than yields from other plots. Plot 11/1(a) gives a marginally higher yield than plot 14/2(d) and plot 11/1(d), where yield appears to be decreasing slightly, reflecting the different treatments as discussed in Section 9.2.1. Assessment of the profitability relationship between two plots leads to similar results, with some differences being caused by the inclusion of inputs. Plot 3(d) is slightly less profitable than plot 14/2(d) but slightly more profitable than plot 11/1(d). Thus the extra inputs on plot 11/1(d) are not justified by the increased yield, although they are justified on plot 14/2(d). Comparison of the profitability of plots 11/1(a) and 11/1(d) shows plot 11/1(a) being almost always better. So although the use of lime implies an additional input cost this is clearly compensated by the increase in yields (Figures 9.6a and b).

### Woburn

The comparison between the two Woburn plots 8a (unlimed) and 8aa (limed) is again an assessment of the use of lime. A comparison of grain and straw yields from the two plots shows clearly higher yields for the plot 8aa. Comparison of overall productivity shows the same pattern, plot 8aa being clearly more profitable (see Figures 9.1 and 9.8).

### Broadbalk

Comparison of the yields from continuous wheat plots 22, 3 and 8 showed much lower yields for plot 3, which had no fertilizer applied (see Figures 9.4b–c). The effect of FYM and medium level fertilizer treatment on yield is not differentiable by the eye. For the more recent treatments, the yield of high level fertilizer (plot 9) was lower than that of FYM + nitrogen (plot 21) which were both introduced in 1968, but slightly higher than that of the medium fertilizer plot 8. Despite the short series of data for the wheat in rotation with potatoes and beans (1968–78), it was thought useful to compare the rotational systems with the continuous wheat. Plot 3 (no fertilizer) and plot 8 (N3PKNaMg–medium fertilizer), were chosen to compare continuous with rotational wheat.

Rather than take the yield of a single crop each year, the rotations were treated as one hectare yielding three crops on one-third hectare each. On a yield/ha basis, this method shows grain yields being higher in rotation, although it must be remembered that this yield applies to only one-third of the area. The difference is clearer for plot 3 than for plot 8 (Figure 9.4.d).

Despite the cost of fertilizers, plot 3 continuous wheat is clearly less profitable than plots 22 and 8 continuous wheat (see Figure 9.10). Furthermore, the FYM plot 22 is more profitable than the medium fertilizer plot 8 despite similar yields. This clearly reflects the low input costs for FYM as opposed to an inorganic fertilizer (although it should be remembered that we had to estimate FYM costs, and a different costing could lead to different conclusions). The profitability relationship between plots 8 and 9 is around 1, indicating that the gain of higher yields on plot 9 gets absorbed by the higher fertilizer cost. Plot 21 (FYM + N2) was more profitable than the medium fertilizer plot 8. Finally the indexes for continuous wheat versus rotational wheat comparisons indicate that the rotational systems are more profitable. In contrast to the yield comparisons the overall comparison takes account of the rotational outputs: wheat grain, straw, beans and potatoes.

## 9.5 TFP INDEXES INCLUDING EXTERNALITIES

A sustainable agricultural system is one that is both economically and ecologically viable (including demand for food and ability to pay for it) in the long term. The final step in using TFP to assess the sustainability of the agricultural system concerns adjustment for externalities. The coventional TFP attempts to account for all marketed outputs and inputs but omits non-marketed goods and services such as environmental quality, although excluding production externalities can overstate (understate) productivity gains, as some resource costs (benefits) are not accounted for.

It is anticipated that the environmental costs of operating a farming system must in future have a major influence on whether it is economically sustainable. Such costs might arise from various sources such as

- soil erosion (from adverse terrain effects or poor husbandry)
- soil exhaustion (from non-replenishment of nutrients)
- biodiversity/aesthetic appearance/wildlife habitats
- pollutant transfer (from nitrate leaching or release of pesticides or herbicides into water courses).

Direct measurement of these effects is difficult, and ambiguous in its interpretation because it is dependent on topography and the farming system practised. For example, Rothamsted soil is a clay loam and erosion is negligible. Furthermore, in the experimental situation, the small scale of the operations means that pollution effects, for example, are difficult to measure, and may not be realistic when scaled up from small plots to more typical field areas.

In attempting to reflect the effects of externalities in the economic indexes that we have used, there are several cases to consider. An externality which is present at a constant level over time will not (by the very nature of the index measures) affect the trend in the index (unless its relative cost increases or decreases over time and a chain-linked index is used), although the value of the output/input ratio will reflect the effect of the externality. On the other hand, if an externality is judged to be increasing or decreasing over time, then this should be visible in both the index and output/input ratio.

Given a quantitative measure (quantity and cost) of an externality effect, it must be added into the index. Since externalities arise as a result of the agricultural system, it is natural to consider externalities as negative outputs. However, in extreme cases this could lead to negative output quantities, in which case indexes could not sensibly be used. For this reason we believe that externalities should be considered as additional costs associated with input factors.

In our study, the main externalities to be considered are crop treatments, i.e. fertilizers, pesticides and herbicides. Soil erosion has not been considered since it is not an important factor at Rothamsted, nor was it at Woburn during the time period considered.

Since no quantification or costing of externalities (or data upon which these could be based) was available to assess the actual sustainability of the Rothamsted experiments, we assessed the behaviour by the approximate method suggested by Steiner and McLaughlin (1992). This method uses proportional costing for crop treatment-related externalities, i.e. the additional externality cost per unit

**Table 9.6** Proportional fertilizer and pesticide cost increases examined

| Fertilizer (% increase) | 0 | 5 | 15 | 30 | 60 | 90 | 120 |
|---|---|---|---|---|---|---|---|
| Pesticides (% increase) | 0 | | 33.3 | 66.7 | 100 | | 200 |

of a treatment with application cost $c$ (per unit) is taken to be $pc$, where $0 < p < 1$. Rather than use a single value for these proportional costs, we have studied a wider range: we have considered potential increases in fertilizer costs and pesticide costs up to 120% and 200% respectively. Although they may be unrealistic, these high levels are used to illustrate what happens to sustainability indexes as we progress from modest proportionate increases to extreme ones, hopefully well beyond the ranges likely to be encountered in practice.

The (externalities-inclusive) indexes are illustrated here for Woburn plot 8a (continuous wheat with medium fertilizer application) and for Broadbalk plot 8 (continuous wheat with medium level fertilizer, pesticide applications since 1960).

In each case output/input ratios and the $A3$ index were calculated using the proportionate increases in costs of fertilizers and of pesticides shown in Table 9.6. The output/input ratios for Broadbalk plot 8 and Woburn plot 8a (Figures 9.12 and 9.13) decrease as increasing externality costs are introduced, as expected.

Figure 9.14 shows the original $A3$ index together with the index adjusted for two sets of externality costs. Although the pattern remains the same, a small change is evident. Since fertilizer is applied at a constant rate throughout the

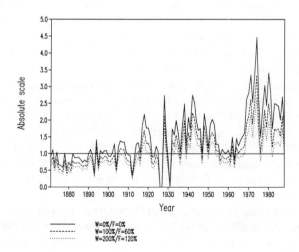

**Figure 9.12** Output/input ratio (productivity) of Broadbalk plot 8 (N3PKNaMg) adjusted for externalities.

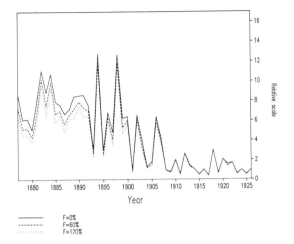

**Figure 9.13** Output/input ratio (productivity) for Woburn plot 8a (N4PKNaMg) adjusted for externalities

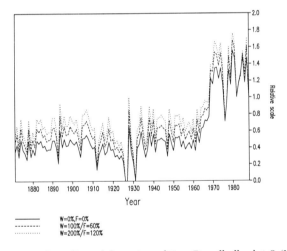

**Figure 9.14** A3 index adjusted for externalities, Broadbalk plot 8 (N3PKNaMg).

experiment, adding a proportional externality cost of fertilizer does not change the index. However, since pesticides were used only after the 1960s, a slight change due to pesticide use is visible. Because adding the pesticide externality cost reduces values of the output/input ratio at the end of the series, the value of the index at earlier points increases, meaning that the overall upward trend is lower since all indexes are constrained to finish at the same point. For similar reasons, the upwards trend in the index decreases as increasing pesticide externality costs are imposed.

## Sustainability—the Rothamsted experience

No pesticides were used on Woburn plot 8a. Fertilizer was used, with the dose being halved in 1882 and again from 1906 onwards. Figure 9.15 shows the $A3$ index for increases in fertilizer costs of 60% and 120%, respectively. No change is visible after 1906. Before 1906, the added externality costs decrease the output/input ratio and the value of the index decreases. As fertilizer externality costs increase, the decrease before 1906 becomes more marked.

It should be noted that even relatively large increases in proportional externality costs of fertilizers and pesticides have modest effects on both the index $A3$ and the output/input ratio.

These results may mean that these levels of externality are compatible with sustainable agriculture. On the other hand they may reflect a problem in this method of accounting for external effects, or in the sensitivity of the $A3$ index. In practice, explicit monetary values must be assigned to the externality costs. Assessment of sustainability may depend heavily on the actual costs used, and the accuracy of these costs therefore becomes very important. At present, externality costs are not well understood and so the sustainability analysis above could be misleading (even ignoring other reservations about the method). An alternative approach to assessing sustainability, which avoids these problems of monetary evaluation (see for example Callicott, 1991), is to use a whole range of system health indicators, where each externality has to satisfy a set of criteria, or bounds, defining acceptable behavior. A system is then sustainable if all the criteria are met. Research would be needed to establish appropriate criteria (in terms of both data and statistical analysis), but the method has the advantage that no monetary externality costs are required, thereby removing a major source of arbitrariness from the analysis. Externalities are then assessed for their direct effects on the system, rather than indirectly via an approximate translation into monetary terms.

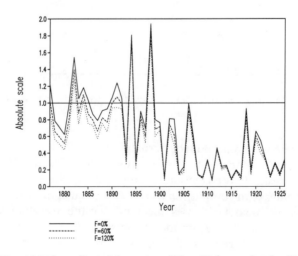

**Figure 9.15** $A3$ index adjusted for externalities, Woburn plot 8a (N4PKNaMg).

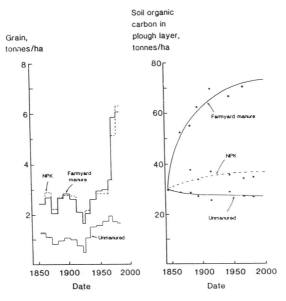

**Figure 9.16** Yields (t/ha) of winter wheat on the unmanured NPK- and FYM-treated soils since 1852 on Broadbalk and the change in soil organic carbon.

Soil quality, a vital component of any system of land use management, is especially important in relation to the sustainability of agricultural systems. Soil quality or fertility depends on complex interactions between the biological, chemical and physical properties of a soil and it can be affected both by agricultural management practices and by inputs from non-agricultural activities. Thus whilst agricultural systems can have external effects, like those of nitrate and pesticide residues, so non-agricultural activities can have effects on soil health indicators and cannot be ignored. Whilst eventually it might be possible to give some appropriate criteria for soil health indicators affecting crop productivity, the complexity of the interactions would probably severely affect their applicability to more than a limited range of soils. Consider three examples of the sorts of factors which should be considered in any total analysis of the system.

(i) Figure 9.16 shows the change in yield on three plots on Broadbalk, the unmanured and those given either NPK fertilizers or farmyard manure (FYM). It also shows the changing levels of organic carbon (surrogate for organic matter) on the three plots. Yields with fertilizers (supplying 144 kg N/ha), and FYM (35 tonnes/ha) have varied little from one another throughout almost 150 years although the organic matter content of FYM-treated soil has increased appreciably. The extra organic matter was not important at these levels of input. In recent years, however, when extra fertilizer N (96 kg/ha) has been given together with FYM on this soil with

**Table 9.7** Maximum yields of winter wheat, grain, t/ha, given by fertilizers, farmyard manure and farmyard manure plus fertilizer N, on Broadbalk.

|  | Cultivar and years grown | | | |
| --- | --- | --- | --- | --- |
|  | Flanders 1979–84 | | Brimstone 1985–90 | |
| Treatment | Continuously | In rotation[†] | Continuously | In rotation[†] |
| NPK[1] | 6.93 | 8.09 | 6.69 | 8.61 |
| FYM[2] | 6.40 | 7.20 | 6.17 | 7.89 |
| FYM+N[3] | 8.13 | 8.52 | 7.92 | 9.36 |

[1] Maximum yield of Flanders given by 192 kg N/ha and of Brimstone by 288 kg N/ha.
[2] 35 t/ha FYM.
[3] 96 kg N/ha.
[†] Wheat grown following two years without cereals to minimise the effects of soil-borne pathogens.

enhanced levels of organic matter, yields of winter wheat have been larger than on any other plot (Table 9.7). The beneficial effect of organic matter is important vis-a-vis our need to enhance yields on a global scale to feed our expanding population.

(ii) Figure 9.17 shows how different treatments have caused the concentration of calcium carbonate to decline in Broadbalk soils since 1843. As expected, decline was rapid in soils given nitrogen as ammonium sulphate (N2). This adverse effect of an agricultural practice on soil quality could be ameliorated by applying ground chalk. But the unmanured soils also lost appreciable

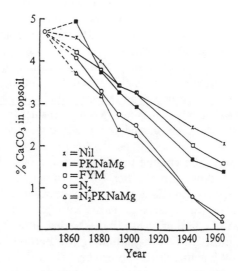

**Figure 9.17** Rates of limestone loss in different plots of the Broadbalk wheat experiment (from Bolton, 1972)

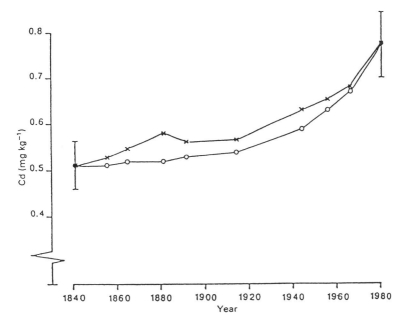

**Figure 9.18** Changes in soil Cd levels in the untreated Broadbalk plot between 1846 and 1980. Measured soil Cd dated (×). Predicted soil Cd concentration (○) based on assumed global emissions of Cd to the atmosphere. Error bars relate to analyses of eight separately digested samples for both 1846 and 1980. All other data points are the mean of duplicate analyses (from Johnston & Jones, 1993)

quantities of calcium carbonate as a result of acidifying inputs from the atmosphere. These externality effects of industrial and non-industrial processes also need to be considered because there is an agricultural cost—the need to replace chalk lost from the soil.

(iii) Figure 9.18 shows the increasing levels of cadmium in the unmanured Broadbalk soil as a result of contamination from atmospheric inputs; again, an externality of an industrial process but one for which agriculture must bear a cost. Little is yet known of the effect of such inputs on soil health, especially on microbial processes occurring in soil. However, we do know from the analysis of herbage on Park Grass that soils must be maintained above pH 6 to appreciably diminish the concentration of cadmium in herbage.

In our view, although a single summary statistic is convenient, in practice a set of indicators would be more informative about different aspects of the system and could be applied more realistically. The TFP index and output/input ratio would be two of these indicators, together with many soil health factors.

This approach also allows some assessment of very nebulous qualities such as landscape quality, for example in its ability to support wildlife populations.

## 9.6 DISCUSSION

During our work on this project, several points have come up for discussion time and time again. They can be summarized as a concern over the method of measuring economic sustainability and a growing appreciation of the problems involved in quantifying externalities. We have shown with our results that two quite different treatment regimes may give rise to similar economic indexes, and that without some overall measure of absolute value as well as the relative value expressed in the index, there is no reliable way of comparing two systems. It is therefore important that an assessment of sustainability is not made by evaluating indexes alone.

Our approach to the inclusion of externalities has been to assess one particular method, namely the inclusion of externality costs as proportional to input costs for a given factor. We have shown that where a treatment regime is reasonably constant an index may show very little change in trend even at very high proportional costs, whereas an absolute measure will reflect the additional costs. Furthermore, although the method of proportional costing may be reasonable for a transient external effect, i.e. where the externality effect disappears before the next season, it is not suitable where an externality effect is in some way cumulative. Thus a more appropriate function must be used once the externality and its magnitude have been identified.

Clearly, we need more understanding of the physical external effects of agriculture before any precise attempts to quantify these effects can be made. For the less tangible effects, such as changes in biodiversity or even visual attractiveness of landscape, quantification is far more difficult. However, it is important that some effort is made to include these factors, as exclusion is equivalent to assigning them zero value.

## ACKNOWLEDGEMENTS

We are grateful for advice and comment during this work to D. S. Jenkinson, J. M. Potts, G. J. S. Ross and A. D. Todd all of Rothamsted Experimental Station.

# PART III
## Review of Findings

# 10
# Incorporating Externality Costs into Productivity Measures: A Case Study using US Agriculture

R. A. Steiner[†], L. McLaughlin[‡], P. Faeth[§] and R. R. Janke[¶]

[†]*Rockefeller Foundation New York, USA*
[‡]*Badi Foundation, Macau*
[§]*World Resources Institute, Washington, USA*
[¶]*Rodale Institute Research Center, Kutztown, USA*

## 10.1 THE CHALLENGE OF EXTERNALITIES

Agricultural productivity, the subject of this book, has been of keen interest to farmers, engineers, economists and policy makers for many years. Productivity is the basic measure of efficiency—productivity growth is recognized as essential to the maintenance of farm profitability and economic health.

Scientists and engineers generally consider productivity in terms of the flow of materials and energy used to produce useful commodities. In agriculture; land, labour, capital and other material inputs have been the key components used for measuring productivity. Official statistics are often reported as 'output per farm labour' or 'output per acre.' Economists measure productivity in a way that relates marketed outputs to purchased inputs. Few attempts have been made to measure and include non-marketed outputs (such as agricultural runoff) or inputs (such as natural resource services), or to determine their significance for

---

*Agricultural Sustainability: Economic, Environmental and Statistical Considerations.*
Edited by V. Barnett, R. Payne and R. Steiner © 1995 John Wiley & Sons Ltd

economic productivity. Even though the economic costs of pollution are known to be considerable and the value of the environment as a repository for production wastes to be high, little consideration has been given to these aspects of *environmental* productivity.

Environmental productivity, however, is an essential element of sustainability. Analyses that purport to consider the sustainability of agricultural systems would clearly be incomplete if natural resource impacts were ignored. The real costs to society of externalities such as groundwater contamination and soil erosion are too important—especially to resource-dependent sectors such as agriculture Faeth *et al.* (1991). This is particularly true for long-term agronomic experiments because of the time frames involved and the cumulative effect of externalities in the environment. Our purpose is not to be prescriptive or comprehensive. Rather we seek to begin exploring ways of measuring externality costs so that they can be incorporated into output/input economic analyses of long term agronomic trials.

Valuing the cost of externalities is a science still in its infancy. Much of the basic knowledge and research methodology needed to understand the effects of externalities remains to be developed. Even when the effects of externalities are reasonably well understood, calculating their costs to society remains a difficult task. Some of these costs can in principle be quantified, but many others involve non-market goods and depend on highly controversial judgments such as the monetary value of human life. Nevertheless, rough estimates of some external costs can be calculated. For some environmental problems, such as those arising from soil erosion, the estimates can be based on measurable damages and can be reported with some confidence. For others, such as the impact of pesticide use on wildlife, the information is much weaker. However, to be truly accurate, externality costs can only be calculated on a location-specific basis—which currently is impossible because of the lack of information.

In this chapter we develop cost ranges for externalities on a national level for the United States because this information is most easily available. We shall systematically examine pesticide, fertilizer and soil erosion externality costs and explore ways in which these can be incorporated into productivity analyses. The methodology would have to be repeated in order to be applicable to other countries.

## 10.2 PESTICIDE EXTERNALITIES

Measuring the externality costs of pesticide use is complicated by the diversity of pesticide products, their environmental impacts, method of application and the specific characteristics of the environment into which they are released. Much of the basic science concerning pesticide effects is still not well understood and the epidemiological evidence needed to infer the health effects of a particular pesticide application reliably is extremely difficult to obtain.

Pesticides enter the environment in numerous ways to cause problems. Pesticides should ideally affect only the target species and quickly degrade, but losses can occur in a number of ways.

(a) Spills, tank washings and storage losses (0–10% of the application).
(b) Losses into the air during application (0–60%, depending on application method).
(c) Leaching through the soil profile into the ground water (<1% of the application).
(d) Surface water runoff (0–0.5% of the application).
(e) Pesticide residues on the crop (<1% of the application) (Madhun and Freed, 1990).

Pesticides also vary in toxicity ranging from highly toxic to relatively benign depending on the species. For example, some compounds that are non-toxic to mammals may be highly toxic to fish. Once the pesticide has moved through the environment the damage that results will depend on the human and non-human life with which it comes in contact. However, because the movement of a pesticide through the environment can be highly stochastic depending on such things as the soil type, temperature and biological activity in the soil, and wind direction at time of application; predicting the externality effects from pesticide application can be very difficult.

However, we can divide externality costs into three categories: regulatory costs, health effects on humans, and environmental costs. Despite the variability of pesticide characteristics and the location-specific nature of their external effects, we can make some *minimum* estimates of a number of the external costs. It is important to note that not all pesticides would contribute to these costs to the same degree; compounds that have higher toxicity, longer half-lives and readily disperse in the environment would contribute more.

## 10.2.1 Regulatory costs

Because of the recognition that a clean environment is important, the US government spends considerable amounts of money on pollution regulation and control. This represents a real and current cost to society that is a result of the use of chemicals. A detailed breakdown of various pollution control costs was recently published in the EPA (1991) report 'Environmental Investments: The Cost of a Clean Environment'. Two basic types of costs, capital and operating costs, were included to represent implementation and compliance costs that result from government environmental regulations.

### Chemical control costs due to pesticides

Pesticide control costs are primarily the result of the pesticide programme implemented under the Federal Insecticide, Fungicide and Rodenticide Act

**Table 10.1** Externality costs due to chemical control regulations for pesticides (FIFRA). Only costs paid by the US public are included. All costs in 1986 dollars

| | 1972 | 1973 | 1974 | 1975 | 1976 | 1977 | 1978 | 1979 | 1980 | 1981 | 1982 | 1983 | 1984 | 1985 | 1986 | 1987 | 1988 | 1989 | 1990 | 1991 |
|---|---|---|---|---|---|---|---|---|---|---|---|---|---|---|---|---|---|---|---|---|
| **Public Costs** | | | | | | | | | | | | | | | | | | | | |
| EPA (a) | 26 | 33 | 35 | 36 | 65 | 64 | 50 | 75 | 69 | 64 | 52 | 48 | 49 | 58 | 53 | 52 | 59 | 111 | 114 | 61 |
| Non-EPA Federal (b) | 0 | 0 | 0 | 0 | 13 | 21 | 16 | 15 | 14 | 11 | 10 | 9 | 9 | 9 | 8 | 8 | 7 | 7 | 7 | 8 |
| Stata Government (c) | 0 | 0 | 0 | 1 | 1 | 3 | 12 | 25 | 23 | 19 | 18 | 17 | 17 | 18 | 17 | 16 | 16 | 24 | 20 | 21 |
| **Private Costs** | | | | | | | | | | | | | | | | | | | | |
| Farm work safety (d) | 0 | 0 | 22 | 20 | 19 | 19 | 19 | 18 | 16 | 15 | 15 | 14 | 15 | 15 | 16 | 15 | 15 | 89 | 154 | 154 |
| Cert/training (e) | 0 | 0 | 0 | 0 | 80 | 67 | 62 | 53 | 44 | 46 | 46 | 46 | 48 | 49 | 52 | 52 | 51 | 54 | 56 | 86 |
| Cancellations (f) | 0 | 33 | 30 | 25 | 45 | 43 | 110 | 161 | 122 | 101 | 73 | 47 | 112 | 40 | 30 | 88 | 67 | 113 | 307 | 427 |
| Total | 26 | 66 | 87 | 82 | 223 | 217 | 269 | 347 | 288 | 256 | 214 | 181 | 250 | 189 | 176 | 231 | 215 | 398 | 658 | 757 |

All figures from EPA publication Environmental Investments: The Cost of a Clean Environment (1991).
(a) p. 6–8, Appendix J, Table J-3, last line: EPA expenditures are associated primarily with pesticide registration (p. 6–3).
(b) p. 6–8, Appendix J, Table J-4, last line: In kind matching funds for certification and training. Also program costs for the National Pesticide Impact Assessment Program.
(c) p. 6–8, Appendix J, Table J-2, last line: Cost associated with certification/training, enforcement and farm worker safety.
(d) Appendix J-2: Based on EPA Office of Pesticide Programs (OPP) staff estimates. Estimates for years 1989 and 1990 reflect new requirements pursuant to a proposed rule to revise worker protection standards.
(e) Appendix J-2: OPP staff estimates.
(f) Appendix J-1B: OPP staff estimates. Although private, these costs are written off as business expenses and thus result in reduced tax revenue.

*Pesticide externalities* 213

(FIFRA). Originally enacted in 1947, FIFRA received a major overhaul in 1972 and has been amended a number of times since, the last in 1988. Pesticide Programme expenditures include those at the Federal and state levels for the registration and re-registration of pesticide active ingredients, certification of pesticide applicators, farm worker safety programs, and enforcement. At the private level (including manufacturers, formulators, distributors and applicators), costs are associated with compliance with FIFRA requirements, including registration-related toxocology and pesticide testing, pesticide disposal, storage and application requirements, and pesticide cancellations and suspensions. (EPA, 1991) these cost figures are listed in Table 10.1 (see also Figure 10.1). The primary private costs are what farmers pay to purchase and apply the pesticides; as these are included in the price, they are not externalities. Other private costs such as farm-level safety and training are not reflected in economic indicators such as prices. Administrative costs of certification and cancellation could be considered externalities as these are handled through federal and state governments.

*Water pollution control costs due to pesticides*

Considerable resources are expended to maintain water quality in US lakes and rivers. Pollution comes from both point (i.e. a factory) and non-point (e.g. agricultural fields) sources. Even though agriculture is the primary contributor to non-point source pollution, it is difficult to attribute regulatory costs associated with non-point source pollution to pesticide and fertilizer use, because most of the costs result from erosion control measures and storm drainage (Crosson and Brubaker, 1982), and probably would be spent no matter what pesticides were used. Thus regulatory costs associated with ensuring safe drinking water, rather than water quality in lakes and streams, are most likely to be directly attributable to pesticide use.

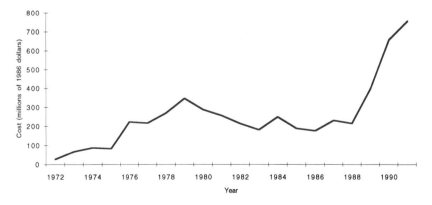

**Figure 10.1** Pesticide regulatory costs annualized at 7% (source: FIFRA)

**Table 10.2** Water pollution control costs for safe drinking water. All costs in millions of 1986 dollars, annualized at 7%

| | 1972 | 1973 | 1974 | 1975 | 1976 | 1977 | 1978 | 1979 | 1980 | 1981 | 1982 | 1983 | 1984 | 1985 | 1986 | 1987 | 1988 | 1989 | 1990 | 1991 |
|---|---|---|---|---|---|---|---|---|---|---|---|---|---|---|---|---|---|---|---|---|
| **Drinking Water** | | | | | | | | | | | | | | | | | | | | |
| Existing Regulations | | | | | | | | | | | | | | | | | | | | |
| EPA(a) | 10 | 9 | 11 | 10 | 19 | 51 | 67 | 85 | 91 | 108 | 95 | 94 | 86 | 87 | 86 | 94 | 95 | 97 | 103 | 108 |
| State Government (b) | | | | | | 10 | 21 | 27 | 40 | 52 | 61 | 65 | 64 | 69 | 82 | 65 | 89 | 97 | 105 | 113 |
| Local Government | 647 | 714 | 807 | 919 | 1042 | 1136 | 1255 | 1397 | 1512 | 1665 | 1799 | 1888 | 1990 | 2131 | 2296 | 2412 | 2430 | 2479 | 2600 | 2723 |
| Private | 145 | 160 | 181 | 206 | 233 | 254 | 281 | 313 | 339 | 373 | 403 | 423 | 446 | 477 | 514 | 540 | 544 | 555 | 582 | 610 |
| New Regulations | | | | | | | | | | | | | | | | | | | | |
| Local Government | | | | | | | | | | | | | | | | | 76 | 154 | 161 | 304 |
| Private | | | | | | | | | | | | | | | | | 17 | 34 | 36 | 68 |
| Total | 802 | 883 | 999 | 1135 | 1294 | 1451 | 1624 | 1822 | 1982 | 2198 | 2358 | 2470 | 2586 | 2764 | 2978 | 3111 | 3251 | 3416 | 3587 | 3926 |
| Pesticide Costs (25%) | 200.5 | 220.8 | 249.8 | 283.8 | 323.5 | 362.8 | 406 | 455.5 | 495.5 | 549.5 | 589.5 | 617.5 | 646.5 | 691 | 744.5 | 777.8 | 812.8 | 854 | 896.8 | 981.5 |

All figures from EPA publication Environmental Investments: The cost of a clean environment (1991).
(a) Table 4–3: Operating cost data from annual Justification of Appropriation Estimates for Committee on Appropriations. Capital costs assumed to be zero.
(b) Appendix F, Table F-12: Capital outlays and O&M expenditures are taken from US Census Government Finance Series.
Data is broken down into three time periods corresponding to the appropriate regulatory phases: Pre-regulatory period prior to 1978, Interim period, 1978–88, and 1986 SDWA Amendments Period covering years 1989–2000.

The Safe Drinking Water Act (SDWA) of 1974 initiated a regulatory programme to develop and enforce uniform national quality standards to assure the safety of public drinking water supplies. The costs of ensuring safe drinking water come from monitoring wells and ground water, treatment costs, and enforcing standards. The US government estimates of the total yearly costs for these activities are listed in Table 10.2. As pesticides are one of the key contaminants of drinking water, a part of these costs are attributable to their use.

The US Environmental Protection Agency has recently completed its five-year National Survey of Pesticides in Drinking Water Wells, finding that at least half of the nation's drinking water wells contain detectable amounts of nitrate, with a small percentage that have concentrations higher than the EPA's regulatory and health-based limits for drinking water (about 1.2% of community water system wells (CWS) and about 2.4% of rural domestic wells) (EPA, 1991).

Researchers at the Economic Research Service of the US Department of Agriculture estimate that first-time monitoring costs range from $0.9 billion to $2.2 billion for households with private wells and are approximately $14 million for community ground water systems. These costs are for monitoring and do not include the cost of remedial action which is often very significant.

Although there is no study showing the percentage of contamination attributable to pesticides, we can make an educated estimate that 25% of the yearly costs associated with safe drinking water can be attributed to pesticides. That is, if pesticides were never used, the cost of ensuring safe drinking water would be at least 25% less. These costs are illustrated in Figure 10.2.

## 10.2.2 Pesticides and human health

There is an increasing amount of evidence that the use of pesticides can have adverse effects on human health (Conway and Pretty, 1991). These can be

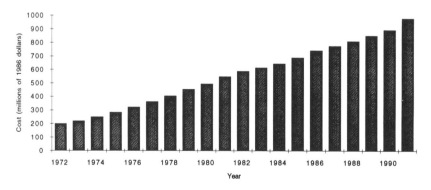

**Figure 10.2** Pollution control costs for safe drinking water (annualized at 7%)

## 216  Externality costs in productivity: US study

**Table 10.3**  Cost of pesticide poisonings in the US. (Estimates by Jerrome Blondell of the EPA, and William Zamula of the CPSC.)

|  | 1985 | 1986 | 1987 | 1988 |
|---|---|---|---|---|
| Total cost per case ($) | 3 404 | 2 350 | 2 868 | 4 166 |
| Number of cases | 15 725 | 13 054 | 17 134 | 14 636 |
| Total national cost estimate ($) | 53 527 945 | 30 677 087 | 49 140 691 | 60 971 663 |

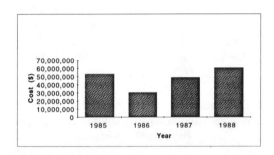

**Figure 10.3**  Cost of pesticide poisonings (reported to hospitals)

separated into acute and chronic effects. The cost of acute effects such as poisonings and dermal exposure are relatively easy to mesure whereas those for chronic effects are extremely difficult to measure.

### The cost of acute health effects of pesticide use

The largest body of information regarding the health effects of pesticides concerns acute poisonings. Massive single exposures, resulting from spillage or deliberate or accidental ingestion, may be fatal (Conway and Pretty, 1991). The most common route of exposure, however, is through the skin. In the United States, children are most likely to suffer acute poisonings because of unsafe storage practices.

The cost of pesticide injuries in the United States was recently estimated by Jerrome Blondell, Health Statistician for the EPA and William Zamula of the Consumer Product Safety Commission, using a database of US Emergency Room records (personal communication). These estimates, reported in Table 10.3, are likely to be low, since many of the likely victims, such as migrant labourers, may not seek medical attention. Another source (Pimentel *et al.*, 1980) gives an estimate of 45,000 cases a year—three times the number found by Blondell and Zamula. If this is a reflection of the under-reporting for 1988, the cost of acute poisonings may be in the range of $60–180 million.

*The cost of chronic health effects of pesticides*

Epidemiological studies needed to determine the chronic health effects of pesticides simply do not exist in the United States. Because of the ubiquitous nature of pesticide use, it is difficult to determine the casual linkages between a human illness and a specific pesticide. There are knowledge gaps regarding both biological and non-biological factors (Rosenstock, 1992). The biological factors include the facts that exposure measures are poor, the symptoms of these effects are not typical of known diseases, and involve chronic diseases with variable and often long latencies. Other factors include inadequate pesticide use data, weak medical and health infrastructures, poor characterization of populations exposed to risks, and economic disincentives for accomplishing such research.

Because of this lack of knowledge and the high cancer incidence among farmers, the National Cancer Institute, the National Institute of Environmental Health and the Environmental Protection Agency are undertaking a five-year epidemiological study on farm chemicals and health which will monitor the lifestyles and work habits of some 100,000 farmers and their families (Hamilton, 1992). Until this study is complete only very crude estimates are possible. By looking at the relationship of other acute and chronic disease costs we can estimate that the value of chronic effects of pesticide use are not less than the minimum costs associated with acute effects ($60 million).

### 10.2.3 Environment externalities of pesticide use

As with other pesticide related externalities, lack of reliable data and complex environmental interactions make accurate evaluation of environmental externalities difficult. Although analysis of potential interactions (for example, see Madhun and Freed, 1990) and of specific incidences of pesticide related damage to the environment is available, comprehensive studies examining the economic costs of environmental externalities are limited. In a 1980 study of the costs of pesticide use, Pimentel *et al.* (1980) identify six categories of 'environmental costs':

1. reduced natural enemies and secondary pest outbreaks;
2. pest resistance;
3. crop and tree loss;
4. bee poisonings;
5. fish and wildlife losses;
6. domesticated animal deaths.

An additional category, biomonitoring of pesticide contamination, should also be added, but this was already considered in the section discussing pesticide regulation.

### Reduction of natural enemy populations, pest resistance, crop and tree loss

Of the above categories, both reduced natural enemies and increased pest resistance are borne by the farmer in the form of crop loss and additional pesticide costs. Although other farmers suffer too, we will assume that they are also using pesticides and therefore these costs will not be considered as externalities.

Category 3, crop and tree loss, refers to loss of production when: (i) pesticides suppress crop growth, development, and yield; (ii) residual herbicides prevent intended crop rotations or inhibit growth of crops that are planted; (iii) pesticides drift from the target crop to adjacent crops, or (iv) excessive residues accumulate on crops and make destruction necessary (Pimentel et al., 1991). Of these production losses, most are borne by the farmer in the form of reduced yield. Pesticide drift occurs with both aerial application and ground application, and can lead to costly investigations and litigation as well as significant crop loss. Exact costs are difficult to obtain, however, because many disputes are settled privately and losses may never be reported to federal or local authorities. Still, the risk of litigation is significant enough to have caused most commercial applicators of pesticides to carry liability insurance. This cost can be assumed to be transferred to the farmer through higher costs for pesticide application services.

Investigations into pesticide related crop loss and legal and administrative costs associated with seizure of crops with excessive pesticide residue are costs of pesticide use borne by the public and are not reflected in farm level cost-benefit analyses. They EPA pesticide programme grants for enforcement were $8.6 million dollars in 1990 and are expected to reach $10.4 million by 1995 (EPA, 1991, Appendix J-3A of Costs of Controlling Pesticides). These figures are included in the regulation cost section, as is the biomonitoring cost.

### Bee poisonings

Bee poisoning represents another environmental cost associated with pesticide use. Honey-bee poisonings from pesticides have been reported since the late 19th century (Morse, 1989). Although all fungicides and most herbicides are essentially non-toxic to bees (Johansen et al., 1983), insecticides can pose a serious threat. The biological interactions are complex. Impact depends on many factors including: bee species; the strength of the bee colony; the residual degradation time of the pesticide, its formulation, and the method of application; the time of application; whether the crop is in bloom; air temperature; and the availability of alternative forage (Johansen and Mayer, 1990). Even when bees are not immediately killed from pesticide contact, colonies can be weakened, resulting in reduced foraging, ineffectual care of the queen or brood, reduced resistance to natural enemies, and subsequent poor survival over winter (Webster and Peng, 1989).

Because honey bees and wild bees are responsible for pollination of many crops, such as alfalfa, almonds, apples, pears, and other fruits, vegtables and oil-seed

**Table 10.4** Pesticide related bee losses

| Impact | Value (million $) |
| --- | --- |
| Colony Loss | 13 |
| Loss of honey & beeswax, partial kills and relocation costs | 25 |
| Potential honey production loss | 27 |
| Bee rental to enhance pollination | 0.5–6 |
| Crop loss due to reduced pollination | 20–200+ |
| Total | 86–272+ |

crops, colony loss or reduction in the bee vitality can result in decreased crop yields due to reduced pollination as well as loses in honey and wax production.

In 1980 the total cost of bee loss was estimated at $135 million (Erickson and Erickson, 1983). However, that estimate only reflects provable incidence of colony loss, not the sublethal effects of exposure to pesticide, relocation costs, or bee rental to replace lost pollination. A more recent estimate places these costs at $13.3 million for colony losses; $25.3 million for honey and wax losses, relocation costs, and partial kills; and $27.0 million for losses of potential honey production in areas where bees cannot be kept because of high pesticide use (Mayer, cited in Pimentel et al., 1991).

In 1989, the value of bee rental was estimated at $2 million (Morse, 1989). More recent estimates place the value at approximately $50–60 million per year (R. A. Morse, Cornell University, personal communication). Not all of that amount can be attributed to loss of pollinators due to pesticides. Estimates range from 3–10% (R. A. Morse, Cornell University, personal communication; Pimentel et al., 1991). The highest cost by far, however, is in the loss of potential crop production. The most recent estimate of crop loss due to bee poisonings is $200 million (Pimentel et al., 1991) assuming a 10% decrease in production. A more conservative estimate would be a decrease of 1%, leading to a 20 million external cost. These crop losses can be considered externalities as they are generally not borne by the farmers who apply the pesticides; rather they are borne by farmers in nearby locations who are growing bee-pollinated crops.

## Fish losses

Pesticide exposure can injure fish populations in a variety of ways. High pesticide concentrations can kill fish outright; low concentrations can kill susceptible fish fry, or they can affect reproduction resulting in fewer viable offspring; and insect populations essential for fish survival can be eliminated or reduced. In addition,

## 220  Externality costs in productivity: US study

contamination fish results in economic losses for the fishing industry, and monitoring of fish contamination increases regulatory costs.

Although fish kills have been monitored since 1960, exact losses due to pesticide exposure are difficult to estimate because details on estimates of the number of fish killed in an incident are often unavailable and even when available often miss fish that have sunk to the bottom or have been washed away (EPA, 1990). Therefore, EPA estimates of 141 million fish killed per year from all causes between 1977 and 1987 are undoubtedly low, as are the estimates of 6–14 million fish killed per year due to pesticide poisoning (EPA, 1990; Pimentel et al., 1991).

Figure 10.4 shows the estimated number of fish reported killed due to pesticides, fertilizers, manure and silage drainage between 1960 and 1987. These numbers include only those from states that reported to the appropriate agency; no estimates are made for those states that did not report losses. The pesticide loss data only reflects losses that can be attributed to on-farm pesticide use that introduced pesticides into water bodies either through runoff or settling from the air. Accidents during transportation of pesticides, improper handling, storage and washing of containers are not included.

It is clear that fish kills due to pesticide use vary dramatically from year to year. During the period shown, no trend is evident, perhaps because the monitoring system is inevitably biased toward massive fish kills. It is reasonable, therefore, to use the average number of fish reported killed as an estimate of minimum economic loss. Between 1961 and 1985, on average 1.74 million fish were killed each year due to on-farm pesticide use. Using the American Fisheries Society valuation of $1.70/fish, fish losses can be valued at $2.89 million per year as a lower limit. This estimate is very conservative as companies have been fined much more per fish for damages due to environmental pollution (Pimentel et al., 1991). A conservative upper limit to economic damage can be calculated by assuming that on average 20% of fish kills are unreported (including both

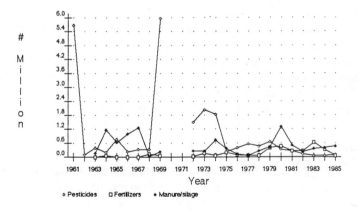

**Figure 10.4**  Fish kills 1961–85

underestimates of numbers killed, cases where the cause is not determined but is in fact due to pesticide use, and states from which no report is received), and by valuing fish at $10 each, (Pimentel *et al.*, 1991). Using those numbers, the maximum economic damage from fish kills would be $21 million.

## Wildlife

Wild birds and mammals can be killed by direct exposure to pesticides as well as by secondary poisonings caused by consuming contaminated prey. In addition, exposure to pesticides, whether directly or indirectly, can reduce survival and reproductive rates even when the dosage is insufficient to cause death. Exposure can also change bird behaviour, causing them to be more vulnerable to predation or to neglect their young (Cox, 1991).

It is difficult to estimate the number of birds and mammals affected, however, because death often does not occur near the site of pesticide exposure, and even in cases where the animals do not travel far, their secretive habits and natural camouflage may make discovery of poisoned animals difficult (Pimentel *et al.*, 1991). In addition, carcasses of poisoned animals may often be consumed by predators before any determination of cause of death can be made (Cox, 1991). persistence of chemicals in the environment can also cause the effects of pesticide use to remain even after the offending pesticide is banned or replaced, further complicating attempts to evaluate externalities (Cox, 1991).

The toxicity of a particular pesticide depends on the formulation used as well as the chemical itself. Granular carbamate insecticides have been found to be particularly lethal to birds, but organophosphates and herbicides have also been found to kill birds in small doses (Cox, 1991). Mineau (1987) estimates that 0.25–8.9 birds/ha are killed by pesticides in the United States each year. Based on estimates of 4.2 birds/ha in US cropland and assuming exposure occurs only on the estimated 160 million hectares of US cropland receiving the most pesticides, Pimentel *et al.*, (1991) calculated that 672 million birds are exposed to pesticides each year. They further estimate that 10% of these birds are killed, giving a total loss of approximately 67 million birds in the United States per year, an estimate on the lower end of the 0.25–8.9 birds killed/ha range given above. No single figure can accurately indicate the value of each bird lost; estimates range from $0.4 for contribution to the economy generated through bird watching to $800 based on the replacement cost of the affected species (Pimentel *et al.*, 1991). Given the range of values between $0.4 and $30, the economic loss due to pesticide poisonings is between $27 million and $2 billion. This estimate is most likely low as it does not take into account either secondary poisonings or decrease in populations due to reduced reproductive and survival rates, nor does it take into account the fact that effects occur disproportionately among species.

Although data related to fish and bird poisonings from pesticides are available, the data related to mammals are sparse. Changes in pesticide damage to birds or other wildlife over time are also very difficult to gauge. Increased awareness of the

**Table 10.5** Pesticide externalities in the US

| Impact | Value (million $) |
|---|---|
| Chemical control costs | 213 |
| Safe drinking water costs | 813 |
| Acute pesticide poisonings | 61–180 |
| Chronic health effects | 61–? |
| Bee-related losses | 86–272 |
| Fish-related losses | 3–21 |
| Other wildlife-related losses | 27–2,000 |
| Total | 1,270–3,560+ |

potential danger to wildlife has resulted in the banning of certain harmful pesticides and in changes in the application rate or method of some of the pesticides commonly used—both of which could be expected to decrease the change of pesticide exposure. On the other hand, lower pesticide dosages have often been achieved by increasing the toxicity of the chemicals applied. Likewise, the decline in wildlife habitat could result in greater exposure to cultivated lands and therefore in increased pesticide exposure. Because of the uncertainties surrounding wildlife injury from pesticide exposure over time, we have not included any estimation of damage to mammals, and we have assumed that the incidence of exposure in birds has remained constant over the period in question.

Since we only have time series data for regulatory costs, we must make certain assumptions about changes in costs over time for the other factors. A very simplistic approach is to extrapolate the other costs by assuming that they were small in 1950 when pesticides first started to be introduced and have increased linearly to today's level. However the analysis is complicated by the fact that pesticides have changed dramatically both in quantity and type and one could argue that early pesticides such as DDT were much more destructive. Nevertheless Figure 10.5 is a reasonable first estimate of the minimum of total externality costs over time.

It should be emphasized that the above figures are quite conservative and do not accurately reflect important costs such as:

- The loss of biological diversity due to pesticide use.
- The cumulative effects of pesticides in the environment.

### 10.2.4 Incorporating pesticide costs into the productivity analysis

The pesticide externality costs developed above represent only those not typically reflected in the prices of inputs and outputs of farming systems. They are also only

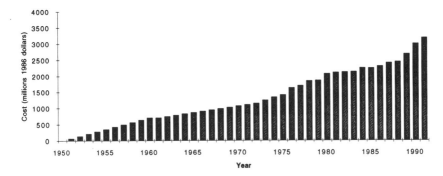

**Figure 10.5** Externality costs of pesticides (1950–90 estimates, United States)

valid on a national level in the US and vary considerably from region to region. As a rough approximation and as an alternative to looking at location-specific effects of pesticides, the overall use and externality costs of pesticides in the United States can be related to each other. The Agricultural Statistics Service estimates that $4 billion were spent on pesticides in 1987 in the US. Using the range of estimates from Table 10.5, minimum externality costs range between 30% and 90%+ of the total expenditure for pesticides. One simple way to include externality costs in output/input economic analyses would be to take a percentage of the price paid for pesticides and count that as the external cost. For example, if the pesticide expenditure is $20 per hectare per years then the range of external costs could be estimated to be between $6 and $18 per hectare per year. This method has several problems, however, and the results should not be used as anything more than a rough indicator. In particular, such an approach ignores the progress that has been made as a result of restrictions and bans on pesticides, as well as the development of more environmental benign compounds, which may, in fact, cost more.

In an example taken from the electric power industry, Repetto (1990) included environmental and health costs in an evaluation of the cost-effectiveness of regulations on particulates, sulphur oxides, and nitrogen oxides. Single factor productivity analysis showed that the kilowatt-hours produced per unit of fuel input and capital input had gone down over time, while labour productivity had gone up a little. A conventional analysis that stopped there would have concluded that regulations had a negative economic impact. However, similar analysis for pollutants showed that the kWhs produced per unit of emissions had gone up by almost 150%. Multifactor productivity measures showed that when the economic benefits of reduced environmental and health benefits were included, productivity increased dramatically over time. When these benefits were excluded, productivity decreased.

A parallel can be drawn from this example to the analysis of interest. The most appropriate way of including pesticide externalities would be to consider the compounds, or classes of compounds used and the rates of application. Where a simple analysis based upon pesticide expenditure would show increasing external costs over time, an analysis that considered the environmental and health impacts of the chemicals used may show a different trend. In the example of Repetto (1990), including emissions as a percentage of fuel costs would have shown little change in environmental and health costs, when in fact there was a significant improvement.

## 10.3 FERTILIZER EXTERNALITIES

Fertilizers are applied to agricultural fields primarily to augment deficiencies in nitrogen, phosphorous, and potassium but increases in agricultural productivity due to fertilizer use are rarely optimized with respect to all the associated costs. Three externalities of fertilizer use are discussed: regulatory costs, the health effects of nitrates in drinking water, and environmental effects of fertilizer use, particularly eutrophication of water bodies caused by phosphorous loading.

### 10.3.1 Fertilizer regulatory costs

Nitrogen may be lost as runoff or leached to infiltrate into the ground-water. Nitrate contamination of ground-water poses a significant health hazard to rural communities who rely on this source of drinking water. Using a computer model to identify relative risk of contamination and carrying out a two-year project in which private and community wells were tested for nitrates, the EPA estimated that 52% of community water system wells contain detectable levels of nitrates and that nationally, 2.4% of rural domestic wells are contaminated above EPA standards (EPA, 1991).

Determining the monitoring costs for nitrates is complicated by the fact that nitrates and pesticides are often found together and testing for one will influence the cost of testing for the other. The average cost of monitoring a private well for nitrates is 11% of the cost of monitoring for pesticides ($813 million). Additionally, 30% of the population with potential ground water contamination is affected by nitrates. Annual monitoring cost due to nitrates, given the above figures, may range from $21–$41 million in 1987. This is roughly between 0.6% and 1.3% of total regulatory costs associated with ensuring safe drinking water. Since the source of nitrates in water can include many factors beside fertilizers, one cannot attribute the entire amount to fertilizers.

### 10.3.2 Fertilizer health costs to society

Health risks associated with fertilizer application are primarily from exposure to nitrates created during the nitrification process. Evidence linking fertilizer application and human health risk is at best limited. While acute poisonings are easy to evaluate, the risks from long-term, low level exposure are difficult to prove conclusively. Ill-effects are likely to show up only after many years and may go undiagnosed, because cause and effect relationships are difficult to prove. Nor, in the case of nitrates, is agriculturally contaminated water the only source of exposure or fertilizers the only source of nitrates in ground-waters. Other sources of ground-water pollution include animal operations and septic fields. Individuals may also be exposed to nitrates through foods, and nitrite levels are naturally high in some water bodies.

Health risks of exposure to nitrates have not been documented adequately. The most serious health hazard, however, is posed by nitrites, also created in the nitrification process. Nitrites may cause 'blue baby syndrome,' known as methemoglobinemia (Phipps et al., 1986). In this syndrome, nitrates are converted to nitrites in the infant's digestive tract and are then absorbed into the bloodstream. There they interact with hemoglobin to produce methemoglobin. Because methemoglobin does not carry oxygen to the cells, the oxygen supply is reduced. In infants the syndrome can be fatal.

Although about 5,000 cases of methemoglobinemia occur each year in the United States, there are no estimates of the approximate costs or number of cases of blue baby syndrome that are due to nitrites in drinking water (Dr. Joe Larry, Center for Disease Control, personal communication.) Nor is any data available for the costs of low level exposure to nitrates.

### 10.3.3 Environmental costs of fertilizers

Eutrophication, a natural process involving growth of aquatic plants, is greatly increased by the presence of fertilizers, particularly phosphorus, in bodies of water. The resulting algal blooms cause the water to become exceedingly turbid. The decay or organic matter in water depletes oxygen levels to such an extent that marine life dies. Other adverse affects of eutrophication include the loss of recreation benefits.

No estimates are available of the recreational costs of eutrophication, but recent monitoring of fish kills by the EPA has included analysis of the numbers of fish killed by eutrophication. In 1986 and 1987, 84,459 and 44,334 fish deaths were attributed to eutrophication. Although eutrophication was not listed in prior EPA analyses of fish kills, in the period from 1961–85, on average 310,000 fish were killed each year by 'fertilizers'. An additional 930,000 fish kills were attributed to manure and silage runoff. Valuing each fish at $1.70 according to the American Fisheries Society valuation as a lower limit and $10/fish as an upper limit, a total annual cost of 2.1–12.4$ million dollars could be attributed to fertilizer effects on fish.

Fertilizer application can also affect other wildlife, which like humans can suffer both acute and chronic effects of nitrate and nitrite poisoning. Livestock

**Table 10.6** Minimum costs of fertilizer externalities (1987)

| Impact | Value ($ millions) |
|---|---|
| Regulatory costs | 11–21 |
| Health effects | 0–? |
| Environmental effects | 2–12 |
| Total | 12–33+ |

kills have also been reported. However, such effects remain isolated incidences with no comprehensive data.

### 10.3.4 Incorporating pesticide costs into the productivity analysis

Given that the total quantity of nitrogen fertilizer used in 1987 was about $2 billion, the externality costs are between 1–2% of the total cost. This is probably too small to make a difference in productivity analysis.

## 10.4 EXTERNALITIES ASSOCIATED WITH SOIL EROSION

### 10.4.1 Off-site costs

Annual off-site damage from soil erosion for each farm production region of the United States has been calculated by the US Department of Agriculture (Ribaudo, 1989). Table 10.7 shows the damage categories and the range of estimates. Table 10.8 shows the estimates for each production region. These estimates vary according to the level of economic activity in a region and the availability and use of surface waters.

This cost figure, however, is at just one point in time and it is clear that erosion costs change over time as population densities, economic activity and surface water use increase. To capture this time trend we deflated the recent damage estimate using US gross national product estimates in current dollars. Figure 10.6 shows an example for the north-eastern section of the country.

Determining erosion rates for each cropping system is not difficult, even though many long-term experiments have not measured soil erosion. Estimates of soil run off for each cropping system can be made using the Universal Soil Loss Equation and standard coefficients, or simulation programs such as EPIC (Erosion/Productivity Impact Calculator). EPIC simulates the physical change in the soil that would occur under different agronomic practices and generates estimates of soil erosion and productivity (Williams, *et al.*, 1989).

**Table 10.7** Annual off-site damage from soil erosion in the United States

| Damage category | Off-site damage (US $ Million)[a] | |
|---|---|---|
| | Best[b] | Range |
| Freshwater recreation | 2404 | 955–7580 |
| Marine recreation | 692 | 497–2772 |
| Water storage | 1260 | 756–1761 |
| Navigation | 866 | 616–1078 |
| Flooding | 1130 | 755–1787 |
| Roadside ditches | 618 | 310–929 |
| Irrigation ditches | 136 | 68–184 |
| Freshwater commercial fishing | 69 | 61–96 |
| Marine commercial fishing | 451 | 443–612 |
| Municipal water treatment | 1114 | 573–1655 |
| Municipal and industrial use | 1382 | 768–1848 |
| Steam power cooling | 28 | 24–39 |
| Total | 10150 | 5826–20341 |

Source: Ribaudo (1989)
[a] Costs updated from 1986 to 1990 dollars using a multiplier of 1.157.
[b] Best estimate is the most likely extent of off-site damage.

Given the erosion rates the off-site damage can be estimated by multiplying by the damage estimate on a tonnage basis for that region. For example, for a continuous corn system in the Northeast, a typical erosion rate may be 20 tons per hectare per year. Using a damage estimate of $8.98 per ton, the off-site costs would be $179.60 per hectare per year. In the TSFP analysis this would appear in the outputs.

### 10.4.2 Positive externalities associated with soil quality enhancement

So far, we have considered the costs associated with farming practices that rely on pesticides, soluble chemical fertilizers, and those that promote soil loss or erosion. There are a number of farming practices, however, that have a beneficial, or net positive effect on the soil, and reduced impact on the environment. These practices are often associated with organic farming or sustainable agriculture, and include:

- rotating crops to increase crop diversity over time;
- intercropping to increase crop diversity through space and keeping the soil covered through the winter months as well as the growing season;
- the use of 'green manure' or legume and grass cover crops that contribute organic matter and biologically available nitrogen to the soil;

**Table 10.8** Off-site damage per ton of soil erosion, by region, in the United States

| Region | Gross erosion (t/yr) Millions | Off-site damage ($/t)[a] Best | Range |
|---|---|---|---|
| Appalachian | 486 | 1.63 | 1.48–2.61 |
| Corn Belt | 967 | 1.33 | 0.65–2.36 |
| Delta | 242 | 2.82 | 1.72–9.40 |
| Lake States | 181 | 4.32 | 2.30–6.92 |
| Mountain | 775 | 1.29 | 0.73–1.99 |
| Northeast | 187 | 8.16 | 4.85–16.25 |
| Northern Plains | 669 | 0.66 | 0.37–2.92 |
| Pacific | 679 | 2.87 | 1.76–4.59 |
| Southeast | 250 | 2.22 | 1.35–3.12 |
| Southern Plains | 490 | 2.33 | 1.33–4.50 |
| Total | 4,925 | 2.06 | 1.19–4.13 |

Source: Ribaudo (1989)
[a] Costs updated from 1986 to 1990 dollars using a multiplier of 1.157.

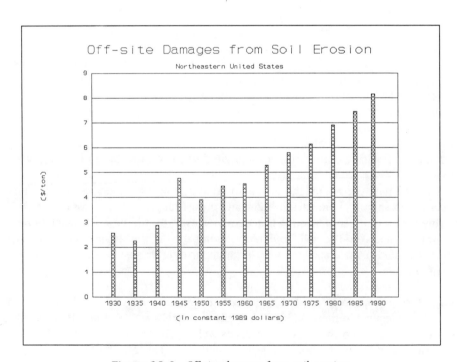

**Figure 10.6** Off-site damage from soil erosion

- the use of composts and manures, which replenish the carbon as well as other nutrients in the soil.

The net results of these practices are measured increases in positive soil properties such as higher organic matter content, enhanced microbial populations and microbial activity, more efficient nutrient cycling and nutrient retention, and more rapid water infiltration and higher water holding capacities in these soils, often referred to as 'tilth'.

The missing piece of information now concerns how to aggregate these separate measurements into an index or other way of quantifying the net positive benefit. Some research is being done to develop a definitive list of important soil attributes to be measured and ways of combining them into a report card, tilth index or soil quality index that could be useful both to farmers wishing to determine whether their practices have increased the value of their land, and policy makers interested in policy tools that create incentives for soil improvement.

The improvement of the world's soils could have significant economic impact on society, since 46% of the world's soils are used for arable crop production. Besides increasing the value of farmland, the tighter nutrient cycling could decrease the amount of nitrate leaching into ground water supplies. Also, the added carbon could tie up significant amounts of carbon dioxide from the atmosphere; which could have an impact on global warming trends.

The long-term studies reported here were not designed explicitly to measure all of the factors associated with soil improvement, but many have measured the soil carbon (organic matter) content since the inception of the experiments. Some of the treatments clearly have depleted the soil carbon, for example continuous monocropping with corn, cotton, or other row crops. Some have maintained the organic matter content by rotating the row crops with soil building crops such as hay or other cover crops, and others have accumulated organic matter through the addition of animal manures.

## 10.5 CONCLUSIONS

Valuing the cost of externalities is still an infant science—much of the basic knowledge and research methodology needed to understand the effects of externalities remains to be developed. As a result, despite their importance in evaluating the sustainability of agricultural systems, externalities continue to be neglected in most productivity analyses. This chapter illustrates the difficulties of measuring externalities and in the context of this analysis offers an initial approach that is both macro and conservative.

As our knowledge in this critical area increases, we will no doubt develop more accurate and justifiable estimates. In the meantime, interim solutions such as this one will have to suffice. Using the inaccuracy of our knowledge as an excuse to ignore externalities would be an unconscionable error—it would effectively assign a value of zero to environment costs—one even the most sceptical critic would have difficulty defending.

We expect the approaches taken here will be revised, refuted and refined—a process we embrace. There is no doubt that improvements are needed, especially when we know that we are failing to include essential elements such as biodiversity and are as yet unable to consider interactive and cumulative effects on the environment. We hope, however, that these efforts bring us closer to including the true cost of externalities in the evaluation of agricultural sustainability.

# 11

# Long-term Experiments and Productivity Indexes to Evaluate the Sustainability of Cropping Systems

K. G. Cassman[†], R. Steiner[‡] and A. E. Johnston[§]

[†]*International Rice Research Institute, Philippines*
[‡]*Rockefeller Foundation, USA*
[§]*Rothamsted Experimental Station, UK*

In Chapter 1, Herdt and Steiner suggest that the key issue concerning sustainability of agricultural systems is 'not whether agricultural productivity is changing. It is whether agricultural productivity gains are occurring at the cost of degradation in the underlying resource base which will eventually result in falling productivity'. Although this issue must be defined with respect to time scale, spatial level and socio-economic environment, the impact of the production system on the natural resources that govern future productivity is fundamental to the viability of all agricultural systems. Based on this perspective, we attempt to summarize the experiences and information from the long-term experiments discussed in previous chapters. Our aim is to consider how to maintain relevance in long-term experiments and the issues they should address, to compare analytical approaches for evaluating productivity trends and their relationship to changes in the natural resource base, and to suggest recommendations for the design, conduct and analysis of long-term experiments.

---

*Agricultural Sustainability: Economic, Environmental and Statistical Considerations.*
Edited by V. Barnett, R. Payne and R. Steiner © 1995 John Wiley & Sons Ltd

## 11.1 MAINTAINING RELEVANCE OF LONG-TERM EXPERIMENTS

Long-term agricultural experiments represent a significant investment of time, labour, capital and land. Like any long-term investment, success depends on: (i) selection of an appropriate asset which will increase in value, (ii) monitoring of asset performance, (iii) anticipation of the dominant factors that determine future asset value and the ability to obtain accurate measurements of these factors, (iv) minimizing transaction costs.

The value of an agricultural system can be described by its contribution to the welfare of farm families and rural communities, by its contribution to regional, national and global food supply and its security, and by its influence on environmental quality. Long-term experiments are therefore most valuable when they target husbandry systems of wide applicability that dominate the landscape and food supply of an agroecological zone. The characteristics of the long-term studies discussed in previous chapters suggest that each experiment was well targeted on cropping systems of regional and global importance (Table 11.1). While this may not be surprising, it is striking that the scientists who initiated these experiments chose treatments that remain relevant today.

The four older experiments established 60–150 years ago included treatments with different sources, rates and balance of applied nutrients as well as crop rotations. Issues of nutrient requirements, nutrient balance and system diversity remain just as important today. Moreover, the cropping system in two of the four oldest experiments continues to be the dominant food production system in surrounding regions. Rain-fed wheat systems dominate in the cool, subhumid temperate zone of northern Europe, and the rain-fed wheat/fallow rotation continues to be the primary cropping system in the cool, semi-arid temperate zone in the Pacific Northwest USA. In contrast, cotton production has declined steadily in the southeastern USA due, in part, to the very problem of acid soil infertility addressed in the Alabama Old Rotation Experiment.

Nutrient requirements to sustain productivity are also the focus in more recent long-term experiments established in developing countries at the beginning of the green revolution in the 1960s and 1970s. Unlike the earlier experiments in temperate environments, crop rotation comparisons are lacking, but the selected rotations have since become the dominant food production system in the surrounding agroecological zone. For example, the rice–wheat system monitored in the Uttar Pradesh long-term experiment established in 1972 now occupies about 12 million ha in the warm, subhumid subtropical zone of northern India, Pakistan and Nepal. By 1990, rice and wheat accounted for about 75% of the national food grain requirement of India (Woodhead *et al.*, 1993). In the warm, humid tropics of south-east Asia, rice is grown on nearly 75% of the arable land that is cropped to all cereals, pulses and oil-seeds (IRRI, 1993).

While most of the original treatments and cropping systems in these long-term experiments have retained their relevance to the agriculture in the surrounding agroecological zone, non-treatment variables have been modified in all of them to reflect technological changes adopted by farmers (Table 11.1). Crop cultivars were changed to utilize the best available genotypes as improved varieties were developed. Mechanization of tillage and harvesting operations was introduced in the older studies to reflect changes in commercial practices and to reduce labour requirements. Pesticides are now used where disease, insect or weed pressure reduce yield or labour productivity. Likewise, blanket rates of non-treatment nutrient inputs applied to all treatments were changed when there was evidence that these nutrients were limiting. Thus, increased rates and split timing of P and K were applied to all treatments in the Alabama Old Rotation Experiment, P rates were increased in the Sanborn Field, and Zn was applied in the IRRI Long-Term Continuous Rice Cropping experiment.

Although potentially useful information can be lost by changing non-treatment management practices, failure to change reduces the value of the experiment as a leading indicator for current husbandry practices. The option of splitting treatment plots to compare the previous practice with the new practice is generally not feasible—either because plot size becomes too small or because it imposes a much larger number of treatments and greater carrying costs. Small plot size can reduce the relevance of long-term experiments due to the overestimation of yield from a small sampling area, excessive soil sampling from a limited area, a greater potential for soil movement between plots and the scale dependence of phenomena such as soil erosion.

The accuracy, quality and relevance of data obtained from long-term experiments are also influenced by the site characteristics, experimental design, sampling intervals and the parameters measured. Treatment replication invariably increases experiment size and cost, and it is possible to detect trends in yield or TFP indexes over long periods of many decades without replication when there is a consistent direction of change as seen in Rothamsted Broadbalk plots (Figures 9.10 and 9.12) and in the Alabama Old Rotation (Figures 4.5 and 4.8). The need for replication to reduce random variation is perhaps more important in detecting trends over shorter time periods of 20–30 years and in making definitive comparisons between time trends in different treatment regimes. Replication to reduce unexplained variation in system performance is also more important in rain-fed systems where drought is a common occurrence, where heterogeneous soil is characteristic of the system under study, and in experiments that include crops that are sensitive to disease and insect damage which tends to be spatially variable.

Even in experiments with adequate replication, the ability to explain productivity trends is often limited by the lack of soil measurements, archived soil samples, data on crop nutrient uptake and removal, and weather data. For example, soil measurement data and archived soils from the Old Rotation Experiment in Alabama are not available before 1950, more than 50 years after the experiment was started. In the IRRI Long-Term Continuous Cropping Experiment, where a yield decline is associated with a decrease in the N

234  LT experiments and indexes for cropping systems

**Table 11.1** Summary of cropping systems, productivity trends, and changes in technology that influenced productivity in the long-term experiments discussed in Chapter 4–9. Productivity trends are based on biological efficiency that is estimated by yield trends with constant input levels, or TFP indexes using constant or indexed prices for output and inputs

| Location, Experiment, Analysis period | Agroecological zone[†] | Cropping system | Major system constraints | Treatment[‡] | Productivity trends Since inception | Productivity trends Last 25 years | Changes in technology[§] |
|---|---|---|---|---|---|---|---|
| Alabama, USA Old Rotation Plots 1896-present | warm, subhumid, subtropics | continuous, annual rain-fed cotton | soil erosion weeds soil crusting insect damage | +PK+L +NPK+L winter leg + L | positive positive positive | negative negative negative | mechanization improved varieties pesticides/defoliants increased PK-rates |
| Rothamsted, UK Broadbalk Plots 1843-present | cool, subhumid temperate | continuous, annual rain-fed wheat | soil acidity weeds | +L +NPK+L +FYM+L | positive positive positive | negative flat flat | mechanization HYVs fungicide/herbicide |
| Missouri, USA Sanborn Field 1950-present | warm, subhumid temperate | rain-fed wheat | soil erosion drought | continuous continuous + NPK rotation + NPK | positive positive positive | flat flat flat | HYVs increased P-rate |
|  |  | rain-fed maize | soil erosion drought | continuous continuous + NPK rotation + NPK | flat flat positive | flat flat flat | improved hybrids increased P-rates |

# Maintaining relevance of long-term experiments    235

| Location / Experiment | Climate[†] | Cropping system | Issues | Treatments[‡] | Trend 1 | Trend 2 | Adaptations |
|---|---|---|---|---|---|---|---|
| Oregon, USA; Residue mgt.; 1931–present | cool, semi-arid temperate | rain-fed wheat/fallow 2 yr rotation | soil erosion, drought | burned straw unburned; +N, unburned; +FYM, unburned | negative, positive, positive, positive | negative, flat, positive, positive | HYVs, increased N-rate |
| IRRI, Philippines; Cont. Cropping; 1968–present | warm, humid tropics | annual triple crop irrigated rice | soil N supply, disease pressure | +PKSZn; +NPKSZn | | negative, negative | pest resistant HYVs, Zn application, fungicide |
| Orissa, India; ICAR-LTF Expt.; 1972–present | warm, subhumid tropics | annual double crop irrigated rice | soil acidity, sandy soil | +NPKS; +NPK+FYM | | negative, flat | improved HYVs |
| Uttar Pradesh, India; ICAR-LTF Expt.; 1972–present | warm, subhumid subtropics | annual irrigated rice–wheat | soil salinity | rice+NPKSZn; rice+NPK+FYM; wheat+NPKSZn; wheat+NPK+FYM | | negative, negative, flat, flat | improved HYVs |
| Punjab, India; ICAR-LTF Expt.; 1972–present | warm, semi-arid subtropics | annual irrigated maize–wheat | soil salinity, sandy soil | maize+NPKZn; maize+NPK+FYM; wheat+NPKZn; wheat+NPK+FYM | | flat, flat, positive, positive | improved hybrids, improved HYVs |

[†] Based on FAO classification.

[‡] A positive (+) sign indicates a treatment with addition of nitrogen (N), phosphorus (P), potassium (K), sulphur (S), zinc (Zn), lime (L), or farm yard manure (FYM). The rotation treatment in the Missouri experiment was a three-year rotation of maize–wheat–clover.

[§] Applied to all treatments.

supplying capacity of the soil/floodwater system, archived soils from the early years of the study are not available to make direct comparisons of soil properties with recent soil samples. On the other hand, the soil archive from the Rothamsted experiments goes back to the mid-1800s and provides a valuable reference for investigating the influence of cropping systems on soil quality and vice versa.

Most long-term experiments include treatments with different nutrient input levels. A crucial issue is whether a given nutrient input regime can maintain soil nutrient stocks and availability for crop uptake. Although soil measurements at regular intervals can detect changes in status, the reasons for such changes cannot be determined without an estimate of the nutrient balance. A complete nutrient balance includes the quantities of applied nutrients, plant nutrient uptake and removal in harvested material, leaching losses, gaseous losses in the case of N, inputs from irrigation and atmosphere, and the net change in soil nutrient level. Precise monitoring of all components of the nutrient balance is generally not feasible in long-term experiments because of cost. In some cases reasonable estimates of the components most difficult to measure, like leaching losses and atmospheric deposition, can be obtained from studies on a similar soil and cropping system that were designed specifically to measure them.

Unlike the difficulties in measurement of nutrient losses and atmospheric inputs, it is relatively straightforward and inexpensive to monitor crop nutrient uptake and removal, and nutrient inputs from irrigation water. These measurements need not be taken in each crop cycle on a yearly basis but rather at regular intervals of perhaps two to four years. For N it is possible to construct a nutrient balance based on the quantity of applied nutrients, measurement of crop nutrient removal and irrigation inputs, estimates of atmospheric inputs available in the literature, and the net change in soil nutrient content over time periods of 10–20 years. Nutrient losses can then be estimated by the quantity of N not accounted for in this budget. For other macronutrients such as K, S, Ca and Mg, changes in soil nutrient supply characteristics are not easily accounted for by mass balance. Changes in soil stocks of these nutrients can be detected by monitoring the readily available nutrient pool using a standard soil-test method that correlates with crop uptake. An increase in the available nutrient status would indicate a positive balance for that nutrient. Nutrient depletion would be indicated by a declining availability index—a non-sustainable situation. When nutrient inputs exceed removal in harvested crop biomass and soil nutrient status decreases, nutrient losses that have not been considered or estimated are indicated, and research can be initiated to identify the factors responsible for these losses.

## 11.2 QUANTITATIVE MEASURES OF SUSTAINABILITY IN LONG-TERM EXPERIMENTS

The exercise of estimating total factor productivity (TFP) in long-term experiments on diverse cropping systems in a wide range of agroecological zones both resolved and complicated the key questions about quantitative measures of sustainability. This experience provides insight into the limitations, appropriate use and potential improvements in the methodology as summarized below.

The TFP indexes appear to be better suited to analysis of productivity trends at the farm, district or regional scale than to long-term experiments. The disadvantage of TFP measures in long-term experiments reflects the tendency to maintain constant input levels and management within a given treatment regime for discreet periods of time. As a result, changes in TFP indexes within periods of relatively constant management are driven primarily by ouput, i.e. economic yield, and yield trends in long-term experiments are thus a reasonable proxy for the trend in TFP. By contrast, farmers continually change management practices and technology (Chapter 5, Table 5.2), and TFP indexes and partial factor productivities can be used to monitor output/input efficiency over time despite changes in husbandry practices (Figures 5.2, 5.3 and 5.4). However, working at the farm, district and regional scale requires data that are robust, with all relevant inputs and outputs monitored and recorded, especially when husbandry systems are changing rapidly. Whilst management of long-term experiments tends to be conservative, Johnston and Powlson (1994) recently suggested a greater need for flexibility towards change, which would enhance the value of the experiments to make them relevant to current farming practice. Where available, comparison of data from the two sources would be of immense value. Rapidly changing indexes in farm-derived data would need careful appraisal of their causes if the indexes from the long-term experiments were comparatively stable.

In practice there appears to be very little difference among the functional forms of the TFP index. Researchers at Rothamsted reported only marginal differences when they compared productivity trends estimated by the Laspeyres, Paasche, Fisher and Tournqvist–Theil indexes for data from the Broadbalk plots (Chapter 9). Despite the theoretical superiority of the more complicated indexes, their results support the assertion that the simplest measure of the TFP, namely the Laspeyres, is preferable because it is both easier to calculate and more intuitively grasped. While the lack of differences between simple and more complex indexes could be attributed to the nature of the data from long-term experiments, there was still no evidence of differences among the indexes during periods when there were rapid and large changes in productivity due to changes in non-treatment technologies.

While TFP indexes evaluate productivity patterns over time on a *relative basis*, profitability measures that aggregate actual output values to input costs are a necessary adjunct. Such measures quantify the actual trend in profit, which is a necessary parameter for determining economic viability. For example, in some of the long-term experiments, control plots without inputs of fertilizer nutrients or manure show a non-negative trend in TFP but do not produce sufficient output to be an economically viable option.

There was a higher degree of variability in most TFP time trends which makes it difficult to detect trends that are statistically significant. Some of this variation might be reduced by adequate treatment replication as discussed earlier. But even in replicated experiments, there will still be variability associated with year-to-year fluctuations in yield. This inherent variability in TFP trends, as well as time series of yield data, emphasizes the need for other quantitative measures in order to make judgements about the sustainability of a cropping system. Monitoring profit is as crucial as the TFP and yield trends for evaluating sustainability, as are changes in soil quality if critical thresholds for the key soil properties that govern crop performance are known.

Although attempts were made to incorporate estimates of externalities in some of the long-term data sets, it is clear that the assumptions and estimates used in this exercise suffer from a lack of knowledge of how to value external effects in TFP measures. Adding a negative output value related to some percentage of an input such as fertilizer N or a pesticide simply shifts the TFP index lower without changing the time trend. In cases where the external effect is cumulative or when it must exceed a specific threshold to cause damage, this approach is clearly not appropriate. Better quantitative methods are required for valuation and precision when estimating externalities. Despite present limitations, TFP indexes should

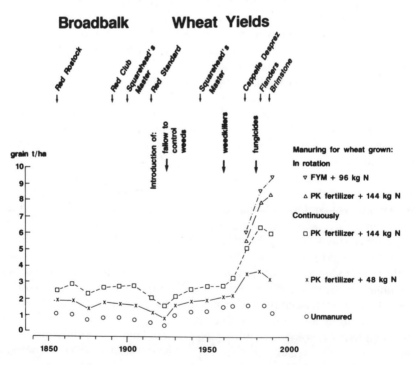

**Figure 11.1** Yields of winter wheat grown on the Broadbalk plots from 1852–90. Treatments are shown at the right, varieties at the top, and changes in general crop management just below the varieties

include at least a rough estimate of externality costs because not to do so implies that environmental damage has no cost at all.

A consistent theme in each of the studies that estimate TFP indexes over long periods is the confounding effect of changes in technology that cause abrupt increases in output. The yield trends and changes in technology used in the Rothamsted Broadbalk plots illustrate this point (Figure 11.1). The problem is illustrated consistently in the studies reported in previous chapters, when traditional varieties with inefficient biomass partitioning are replaced by modern, high yielding varieties with a high harvest index. In the Rothamsted experiment, for example, there is a flat or negative trend in the TFP index in the period from the late 1870s to the 1950s in all treatments, and there has been a rapid upward movement to a higher yield plateau since the introduction of high yielding varieties in the 1960s (Figure 9.9).

Similar trend reversals are also evident in the biological sustainability of rain-fed wheat systems in Oregon due to the switch to high yielding varieties and higher N-rates in 1967 (Figures 6.5 and 6.6). In the Old Rotation Experiment in Alabama, a number of technological changes were made in a relatively short period of 15 years from 1945–60 including mechanical harvesting versus hand-picking of cotton, improved timing of fertilizer application, improved varieties and better insect control with modern insecticides. Taken together, these technological changes resulted in a large yield increase (Figure 4.5) and a twofold increase in TFP (Figure 4.8). Since 1960, after this rapid increase, both yield and TFP index trends have turned mostly negative.

Where abrupt changes in husbandry practices cause a marked discontinuity in productivity trends, it seems preferable to divide time series analyses into periods that separate the distinct departures from previous trends. Such an approach was taken in the Oregon study and provides a clear distinction in the TFP trends before and after adoption of high yielding wheat varieties and increased N fertilizer rates in this long-term experiment.

## 11.3 SYSTEM PERFORMANCE AND SOIL QUALITY

The dominant natural resources of an agroecosystem are soil and water, and both have a large impact on productivity. Although the specific attributes that contribute most to soil quality may differ in different cropping systems and on different soils, for a specified system and soil type quality can be quantified by a subset of physical, chemical and biological properties with the largest influence on crop yields and yield stability.

While quantitative measures of TFP can determine the economic viability and output/input efficiency of a cropping system, they cannot directly detect differences in soil quality. In long-term experiments where input levels and crop management do not change and the relative costs of inputs and outputs are held constant, a negative trend in TFP must be associated with a decrease in soil quality assuming no change in climate. Figure 9.15 illustrates this point for the effect of increasing soil acidity on wheat yields. Thus, monitoring output/input efficiency or even partial factor productivity for inputs with the largest influence on yield provides an indicator for the direction of change in soil quality. Knowing that there has been an increasing or decreasing trend in TFP or partial factor productivity can help focus research on identifying soil properties that have been altered in relation to effects on yield and input use efficiency.

A technological change in crop management can change the relationship between productivity and soil quality, creating a discontinuity. Such a discontinuity makes it necessary to monitor soil properties that are the primary determinants of system performance to predict whether a change in technology contributes to greater productivity without a negative impact on soil quality. In some cases, underlying trends of soil degradation may continue although yield and TFP increase, and soil degradation can be concealed for even longer periods by further improvements in varieties and crop management practices.

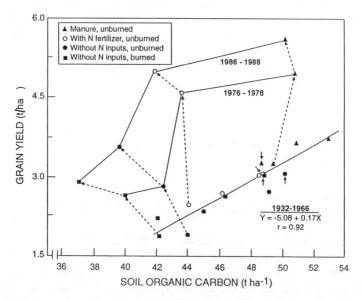

**Figure 11.2** Wheat yield in relation to soil organic carbon content of the 0–30 cm soil layer in the wheat/fallow system of the Residue Management Experiment in Oregon (Chapter 6). Arrows indicate initial conditions in 1931–33. The regression line is for the period from 1931–66 when traditional varieties were used. Modern high yielding semi-dwarf varieties have been used since 1967. Yields are the three-year average beginning with the year of measurement of soil carbon in 1931, 1941, 1951, 1964, 1975, and 1986 (Table 6.3). Dashed lines connect data points of the same treatments for the 1964–66, 1976–78, and 1986–88 periods

This disruption is demonstrated by the relationship between grain yield and total soil organic carbon (SOC) in the top 30 cm of the Oregon Residue Management Experiment (Figure 11.2). Modern high yielding varieties replaced traditional varieties in 1967, and N fertilizer rates were increased in several treatments to support the higher yield potential. In the previous 35 years from 1932–66, a period of relatively constant technology and traditional varieties, there was a close linear relationship between grain yield and SOC: both yields and SOC decreased in treatments without manure during this period. This relationship was fundamentally altered after the introduction of high yielding varieties so that the grain yield was much larger on soil with the same or lower SOC in 1976–78. Yields were increased further in 1986–88 due to continued introduction of varieties with higher harvest index and greater disease resistance, although SOC in treatments without manure continued to decline (Figure 11.2). Part of the decrease in SOC reflects the smaller amount of recycled crop residues as a result of using semi-dwarf varieties with a higher grain/straw ratio.

Soil organic carbon also continues to decline in the unburned +N treatment even though the N rate was increased from 34 to 90 kg N/ha in tandem with the switch to modern varieties in 1967. At issue, therefore, is whether the marked increase in yield and TFP from the introduction of modern varieties and greater N inputs can be sustained in this system as SOC continues to decline. Analysis of the TFP time series since 1967 indicates flat trends in treatments without inputs of fertilizer N or manure, while TFP increases slightly in treatments with fertilizer N or manure (Table 11.1). Because SOC has a large influence on soil N supply and yield in this semi-arid environment, it is likely that both TFP and yield trends will at some point become negative in treatments without manure unless the decline in SOC is arrested. Moreover, although the yield increase since 1967 is impressive in treatments with addition of fertilizer N or manure, the increase has not been large enough to offset decreasing wheat prices and thus economic viability appears to be declining (see Chapter 6).

## 11.4 RECOMMENDATIONS FOR CONDUCTING LONG-TERM EXPERIMENTS AND ANALYSIS OF PRODUCTIVITY TRENDS

The initial design and layout of long-term experiments is crucial to their future relevance. Treatments should focus on those aspects of long-term system performance, such as cropping diversity in time and space, macronutrient inputs and balance, and soil and water conservation, which ultimately govern biophysical sustainability. Testing single factor component technologies on topics that do not threaten long-term viability should be avoided because they are more efficiently addressed in studies of shorter duration. For example, it can be argued that micronutrient deficiencies are more a problem of diagnosis than a threat to sustainability because the cost and quantity of inputs required for correction are small.

The experimental design should be relatively simple, the number of treatments should not be large, and individual plots should be of sufficient size to allow a relevant scale for crop management and accurate yield estimates and to avoid excessive removal of plant and soil materials in monitoring treatment performance over time. Establishing additional 'control' treatment plots without nutrient inputs or with the standard recommended practices at the time of initiating the experiment is useful for later conversion to treatments that incorporate new technologies or crop rotations as agricultural technology advances.

Changes in non-treatment management practices and in the imposed treatments themselves are often necessary to maintain the relevance of long-term experiments. Such changes should not be made capriciously and splitting plots to compare 'old' versus 'new' component technologies should be avoided unless plot size is large and the number of treatments remains manageable.

As in all good experiments, treatments should be replicated in an appropriate experimental design, and the field should be as uniform as possible to avoid the confounding influence of different soil types. If erosion is expected, plot layout and size should be designed to anticipate the direction of soil movement to minimize potential contamination between plots. As discussed in Chapter 2, a minimum data set should be taken at regular intervals and soil samples should be archived for future reference. When crop rotations are included as treatments, a sufficient number of plots should be established so that all crops in a specific rotation can be grown in each year. The requirements for, and benefits of long-term experiments have been recently reviewed (Johnston and Powlson, 1994; Powlson and Johnston, 1994) based on the experience of the Rothamsted experiments since 1843.

In long-term experiments, TFP indexes mostly track yield trends because inputs and management are held relatively constant within a given treatment regime. Thus, TFP as a measure of sustainability may be more appropriate for monitoring and comparing system performance at the farm or regional level (but see Section 11.2). Partial factor productivities for treatment inputs such as inorganic fertilizer or organic nutrient sources can be just as useful as a complete TFP index in monitoring productivity trends in long-term experiments. For example, partial factor productivity for N inputs can be calculated as grain yield divided by the quantity of applied N as inorganic fertilizer or manure (kg grain/kg applied N). Likewise, water use efficiency (kg grain/mm rainfall) is a useful indicator of biological performance over time in experiments that focus on cropping systems in rain-fed environments where drought is a constraint.

Although changes in technology that result directly in greater economic yield and increased output/input efficiency can be accounted for in TFP measures, they can cause a discontinuity in the relationship between economic yield and the quality of the soil resource. In some cases these changes mask continuing soil degradation. Therefore, productivity indexes and yield trends alone are not reliable indicators of the direction of change in the natural resource base on which agriculture depends, namely, fertile soil and an adequate water supply of good quality. Evaluating sustainability also requires direct monitoring of soil properties, and rudimentary nutrient and water balances. Rapid advances in the development of simulation models that predict crop dry matter accumulation, yield formation, nutrient and water requirements now provide powerful tools for estimating yield potential and input use efficiency at potential yield levels (France and Thornley, 1984; van Keulen and Seligman, 1987; Penning de Vries et al., 1989). These predictions can be used as biological efficiency 'yardsticks' for comparison with actual crop performance in long-term experiments, and for interpreting yield fluctuations due to year-to-year variability in climate.

## 11.5 CONCLUDING REMARKS

The value of long-term experiments is ultimately determined by their ability to generate knowledge which might improve the biological and economic performance of important crop production systems. They also serve as an 'early warning system' to detect problems that threaten future productivity. While aggregate data on trends in crop yields and productivity indexes from district, regional and national levels are also important components of this warning system, these data tend to be lagging indicators which are confounded by the rapid changes in technology and large fluctuations in input rates used by farmers. In contrast, well managed long-term experiments are leading indicators of sustainability and provide the opportunity for scientists to investigate cause and effect relationships that govern productivity trends before they are noticed in farmers' fields.

Examples of this early warning capability are the long-term experiments on rice–rice systems in India (Chapter 8) and the Philippines (Chapter 5), and the rice–wheat systems in India (Chapter 8) which indicate significant yield declines despite the continued increase in yield achieved by farmers based on national aggregate data in both countries. Research is now underway to determine the causes of declining yields in these experiments, and to establish whether similar processes are occurring in farmers' fields–trends which may be hidden by rapid advances in the technology used by farmers. Similarly, although wheat yields continue to increase in the Pacific Northwest USA due to improved varieties and crop management practices, data from the long-term Residue Management Experiment in Oregon (Chapter 6) shows that soil organic matter continues to decline. At some point, this reduction in soil quality is likely to theaten the productive capacity and economic viability of this system.

Although negative yield trends are shocking, the lack of positive TFP trends since the 'quantum leap' of the green revolution is also a concern. In most of the long-term experiments analysed in the preceding chapters, TFP indexes based on constant prices for inputs and outputs, or fixed relative price ratios for inputs and outputs, have remained flat or negative over the past 25 years (Table 11.1). In all of these studies, crop varieties are replaced by more recent releases on a regular basis and other aspects of crop management are improved to track the changes occurring in agriculture at large. Where negative trends occur in TFP, there is no evidence that the yield potential of varieties used in the past 25 years has decreased, so declining productivity must reflect degradation in soil quality. Flat TFP indexes suggest stagnant production efficiency, which raises the question of whether this will soon confront farmers in the major food production systems worldwide. If this occurs, the ability to feed an increasing world population could be in doubt.

The limitations of data from long-term experiments must also be recognized. In some experiments plot size is too small for accurate estimates of yield or there is insufficient replication for detecting significant trends over periods of less than two to three decades. Furthermore, research stations are sometimes not representative of the surrounding agriculture because they are established at favourable locations with better quality soil or less slope. Where this is the case, productivity trends in long-term studies may not be comparable to actual trends occurring in farmers' fields.

In contrast to their value as leading productivity indicators, long-term experiments may not be the most appropriate vehicle for assessing negative externalities associated with farming practices. To some extent this reflects the scale dependence of measurements to quantify externalities such as runoff, groundwater pollution from leaching of nitrate or pesticide, or emission of greenhouse gases. It also reflects our inability to estimate the present and future costs of these external effects and the lack of appropriate methods to incorporate these estimates in TFP indexes. These issues are discussed in more detail in Chapter 10, and their resolution remains a critical challenge to scientists concerned with evaluating the sustainability of agricultural systems.

# References

Numbers after the references indicate the page(s) on which it can be found.

Alabama Cooperative Extension Service (1983) *Handbook of Alabama Agriculture*, Circular EX-2, Alabama Cooperative Extension Service, Auburn University, Alabama. 49
Alabama Crop and Livestock Reporting Service (July, 1959; July, 1960) *Alabama Agricultural Statistics*, Alabama Department of Agriculture and Industries, Montgomery, Alabama. 42
Altieri, M. (1986) *Agroecology: The Scientific Basis of Alternative Agriculture*, Westview Press, Boulder Colorado. 12
Anders, M. M. (1990) Sustainable crop production in the Semiarid tropics, *Proc. Ist. Int. Symp. Nat. Resource management for sustainable agriculture*, **1**, 337–447. 134
Anon. (1856–1900) *The Gardeners Chronicle*, London. 183
Anon. (1874–1901) Market Intelligence, Review of the British Grain Trade, *The Agricultural Gazette*, London. 183, 186
Anon. (1888) *Anouncement*, Missouri Agricultural Experiment Station Bulletin 1. 111
Anon. (1991) *Rothamsted Experimental Station, Guide to the Classical Experiments*, AFRC Institute of Arable Crops Research, Harpenden. 173
Ball, V. E. (1985) Output, Input, and Productivity Measurement in U.S. Agriculture 1948–1979. *American Journal of Agricultural Economics*, **67**, 478–486. 48
Barker, R., Herdt, R. W. and Rose, B. (1985) *The Rice Economy of Asia*, Published by Resources for the Future, Inc., Washington, D.C. with the International Rice Research Institute, Los Baños, Philippines. 67
Barber, S. A. (1971) Soybeans do respond to fertilizer in a rotation. *Better crops with plant food*, 55(2), 9. 150
Barnett, V. (1994) Statistics and the Long-term Experiments: Past achievements and future challenges, in Leigh, R. A. and Johnston, A. E. (eds) (1994) *Long-term Experiments in Agricultural and Ecological Sciences*, CABI International, Wallingford. 17
Barnett, V., Landau, S. and Welham, S. J. (1994) Measuring Sustainability, *Journal of Environmental Statistics*, **1**, 21–36. 36, 37
Bartlett, R. J. and Reigo, D. C. (1972) Effect of chelation on the toxicity of aluminium, *Plant Soil*, **37**, 419–423. 156
Benner, R., MacCubbin, A. E. and Holson, R. E. (1984) Anaerobic biodegradation of the lignin and polysaccharide of lignocellulose and synthetic lignin by sediment microflora, *Appl. Environ. Microbiol.*, **47**, 988–1004. 77
Bolton, J. (1972) Changes in magnesium and calcium in soils of the Broadbalk wheat experiment at Rothamsted from 1856 to 1966. *J. Agric. Sci.*, Cambridge, **79**, 217–233. 204
Boyle, G. E. (1988) The Economic Theory of Index Numbers: Empirical Tests for Volume Indices of Agricultural Output, *Irish Journal of Agricultural Economics and Rural Soc.*, **13**, 1–20. 48
Bremner, J. M. (1965) Total nitrogen, in Black, C. A. (ed) (1965) *Methods of Soil Analysis: Part II*, Am. Soc. Agron., Madison, WI. 90

Broomhall, W. M. (1901, 1903, 1904, 1905, 1907, 1909) *The Country Gentlemen's Estate Book, Yearbook of The Country Gentlemen's Association Ltd*, The Country Gentlemen's Association Ltd, London. 180, 183

Brown, W. S. (1950) *Georgia Agricultural Handbook*, The University System of Georgia, Athens, USA. 51

Brown, J. R. and Rodriguez, R. R. (1983) *Soil Testing: A Guide for Conducting Soil Tests in Missouri*, Extension Circular 923 (revised) Missouri Cooperative Extension Service, Columbia, MO. 115

Buchholz, D. D. (1983) *Soil Test Interpretations and Recommendations Handbook*, Mimeograph, Department of Agronomy, University of Missouri, Columbia, MO. 114

Busacca, A. J., McCool, D. K., Papendick, R. I. and Young D. L. (1985) Dynamic impacts of erosion processes on productivity of soils in the Palouse, in McCool, D. K. (ed) (1985) *Erosion and Soil Productivity*, ASAE Publ. 8–85, Am. Soc. Agric. Engr., St. Joseph, MI. 56, 97

Cabrera, P. and Talibudeen, O. (1977) Effect of soil pH and organic matter on labile Al in soils under permanent grass. *J. Soil Sci.*, **28**, 259–270. 156

Callicott, B. (1991) The Concept of Ecosystem Health. Paper presented at the Hastings Centre workshop on *'The Idea of Nature'*, Nov 21–22, 1991. 202

Capalbo, S. M. and Antle, J. M. (1988) *Agricultural Productivity: Measurement and Explanation*, Resources for the Future, Washington, DC. 25

Cassman, K. G., Kropff, M. J., Gaunt, J. and Peng, S. (1993) Nitrogen use efficiency of rice reconsidered: What are the key constraints? *Plant and Soil*, **155/156**, 359–362. 82

Cayton, M. T. C. (1985) 'Boron toxicity in rice, *IRRI Research Paper Series, number 113*, International Rice Research Institute, Los Baños, Philippines. 73

Christensen, L. R. (1975) Concepts and Measurement of Agricultural Productivity, *American Journal of Agricultural Economics*, **75**, 910–915. 48

Christopher, R. C. and Roy. K. B. (1945) Cotton-Hog Farming on the Sand Mountain, *Circular 91*, Agricultural Experiment Station of the Alabama Polytechnic Institute, Auburn, Alabama. 49

Collins, H. P., Rasmussen, P. E. and Douglas, C. L., jr. (1992) Crop rotation and residue management effects on soil carbon and microbial dynamics, *Soil Sci. Soc. Am J.*, **56**, 783–788. 92

Conway, G. R. and Barbier, E. (1990) *After the Green Revolution—Sustainable Agriculture for Development*, Earthscan, London. 3

Conway, G. and Pretty, J. (1991) *Unwelcome harvest: Agriculture and Pollution*, Earthscan Publications, London. 215, 216

Cooke, S. C. and Sundquist, W. B. (1989) Cost Efficiency in U.S. Corn Production. *American Journal of Agricultural Economics*, **71**, 1003–1010. 48, 57

Cooke, S. C. and Sundquist, W. B. (1991) Measuring and explaining the decline in U.S. Cotton Productivity Growth, *Southern J. of Agricultural Economics*, **23**, 105–120. 48

Cox, C. (1991) Pesticides and birds: from DDT to today’ poisons, *Journal of Pesticide Reform*, **11**(4), 2–6. 221

Crosson, P. R. and Brubaker, S. (1982) *Resource and Environmental Effects of U.S. Agriculture*, Resources for the Future, Washington. 213

CSO (1935–1990) *Annual Abstract of Statistics*, HMSO, London. 183

CSO (1964–1992) *Monthly Digest of Statistics*, HMSO, London. 185

Dalrymple, D. G. (1986) *Development and Spread of High-Yielding Rice Varieties in Developing Countries*, Metrotec, Inc., Washington, DC. 65

Davis, F. L. (1949) The Old Rotation at Auburn, Alabama. *Better Crops with Plant Food*, Reprint DD-8-49. Am. Potash Institute, Inc. Washington, DC. 45

Davis, P. O. (1939) *Handbook of Alabama Agriculture*, Alabama Polytechnic Institute Extension Service, Auburn, Alabama. 51

## References

De Datta, S. K. (1986) Improving nitrogen fertilizer efficiency in lowland rice in tropical Asia, *Fertilizer Research*, **9**, 171–186. 80

Dei, Y. and Yamasaki, S. (1979) Effect of water and crop management on the nitrogen-supplying capacity of paddy soils, in Nitrogen and rice, pp. 451–463, International Rice Research Institute, Los Baños, Philippines. 77

Diewert, W. E. (1976) Exact and Superlative Index Numbers, *Journal of Econometrics*, **4**, 115–145. 29

Diewert, W. E. (1989) *The Measurement of Productivity*, Discussion paper 89–04, Department of Economics, University of British Columbia. 31

Dyke, G. V., George, B. J., Johnston, A. E., Poulton, P. R. and Todd, A. D. (1982) The Broadbalk Wheat Experiment 1968–78: Yields and Plant Nutrients in Crops Grown Continuously and in Rotation, *Rothamsted Experimental Station, Report for 1982, Part 2*, 5–9. 177

EC Commission (1991) The Development and Future of the Common Agricultural Policy—Proposals of the Commission, *Com (1991) 258 final*, Brussels. 171

Ehui, S. K. and Spencer, D. S. C. (1993) Measuring the sustainability and economic viability of tropical farming systems: a model from sub-Saharan Africa. *Agricultural Economics*, **9**, 279–296. 3

Environmental Protection Agency (EPA) (1990) *Environmental Damages to Fisheries*, Report of the EPA to the Congress of the U.S., EPA, Washington, DC. 220

Environmental Protection Agency (EPA) (1991) *Environmental Investments: the Cost of a Clean Environment*, Report of the EPA to the Congress of the U.S., Island Press, Washington, DC. 211, 213, 214, 218, 224

Erickson, E. H. and Erickson, B. J. (1983) Honey bees and pesticides: plain talk about the past and present, *American Bee Journal*, **123**(10), 724–730. 219

Evans, W. C. (1977) Biochemistry of the bacterial catabolism of aromatic compounds in anaerobic environments, *Nature*, **270**, 17–22. 77

Faeth, P. (ed.) (1993) *Agricultural Policy and Sustainability: Case Studies from India, Chile and the U.S*, World Resources Institute, Washington DC. 4

Faeth, P., Repetto, R., Kroll, K., Dai, Q, and Helmers, G. (1991) *Paving the farm bill: U.S. Agricultural Policy and the Transition to Sustainable Agriculture*, World Resources Institute, Washington, DC. 210

FAO (1988) Aspects of FAO policies, programmes, budget and activities aimed at contributing to sustainable development. *FAO Council paper* CL. 94/6. 133

Flinn, J. C. and De Datta, S. K. (1984) Trends in irrigated-rice yields under intensive cropping at Philippine research stations, *Field Crops Res.*, **9**, 1–15. 75

Flinn, J. C. De Datta, S. K. and Labadan, E. (1982) An analysis of long-term rice yields in a wetland soil, *Field Crops Res.*, **5**, 201–216. 73, 74, 75

Fox, R. L. (1976) Sulphur and nitrogen requirement of sugarcane, *Agron. J.*, **68**, 891–896. 150, 155

Fox, R. L., Kang, B. T. and Nangju, D. (1977) Sulphur requirement of cowpea and implications for production in the tropics, *Agron. J.*, **69**, 201–205. 150, 155

France, J. and Thornley, J. H. M. (1984) *Mathematical Models in Agriculture*, Butterworths & Co., London. 243

Freeman, O. (1990). Ending hunger and protecting environment are equally important, challenging goals, *Better cropswinter (1989–90)*, 28–31. 133

Frye, W. W. and Thomas, G. W. (1991) Management of Long-Term Field Experiments, *Agronomy Journal*, **83**, 38–44. 17

Garman, C. G. (1932) *Factors Related to Income and Costs of Production on Farms in Marshall and DeKalb Counties, Alabama 1927–1929*, Bulletin 236, Agricultural Experiment Station of the Alabama Polytechnic Institute, Auburn, Alabama. 51

Gollop, F. M. and Jorgenson, D. W. (1980) U.S. Productivity Growth by Industry, 1947–

1973, in Kenricks, J. W. and Vaccara, B. N. (eds) (1980) *New Developments in Productivity Measurement and Analysis*, University of Chicago Press for the N.B.E.R., Chicago, IL.   130

Graham-Tomasi, T. (1991) Sustainability: Concepts and Implications for Agricultural Research Policy, Ch. 3 in Pardy, P. G., Roseboom, J. and Anderson, J. R. (eds) (1991) *Agricultural Research Policy: International Quantitative Perspectives*, Cambridge University Press, Cambridge, UK.   23

Grimme, H. and Von Braunschweig, L. C. (1974) Interaction of K concentration in the soil solution and soil water content on K diffusion, *Z. Pfl. Ernahr. Bodenk*, **137**, 147.   150

Hamilton, D. P. (ed) (1992) Science Scope Tracing disease down on the farm, *Science*, **256**, 299.   217

Hart, R. and Sands, M. (1991) *Sustainable Land-Use Systems Research and Development*, Association for Farming Systems Research Extension Newsletter, Vol. 2, Num. 1.   3

Hastie, T. J. and Tibshirani, R. J. (1990) *Generalized Additive Models*, Chapman & Hall, London.   36

Hayaishi, O. and Nozaki, M. (1969) Nature and mechanisms of oxygenases, *Science*, **164**, 389–396.   77

Herdt, R. W. and Lynam, J. K. (1992) Sustainable Development and the Changing Needs of International Agricultural Research, in Lee, D. R., Kearl, S., Uphoff, N. (eds) (1992) *Assessing the Importance of International Agricultural Research for Sustainable Development*, Symposium proceedings at Cornell University.   10

Herdt, R. W. (1988) Increasing crop yields in developing countries. Paper presented to the *1988 meeting of the American Agricultural Economics Association*, 30 July 3–August 1988, Knoxville, Tennessee.   67

Herdt, R. W. and Capule, C. (1983) *Adoption, spread and production impact of modern rice varieties in Asia*, International Rice Research Institute, Los Baños, Philippines.   65

Hinds, W. E. and Thomas, E. L. (1920) *Poisoning the Boll Weevil*. Bulletin No. 212, Alabama Agricultural Experiment Station of the Alabama Polytechnic Institute, Auburn.   49

Hinman, H. R., Engle, C. F., Erickson, D. H. and Willet, G. S. (1981) *Cost comparisons for alternative tillage practices in western Whitman County*, Ext. Bul. 0840. Wash. State Univ. Pullman, WA.   94

Horward, A. (1925) The origin of alkali land. *Agric. J., India*, **20**, 261.   151, 156

Ingram, B. K. and White, M. (1974) *Indexes of Prices Received by Alabama Farmers. Revised 1959–1973*, Bulletin 456, Agricultural Experiment Station, Auburn University, Alabama.   51

IRRI (1993) *IRRI Rice Almanac*, International Rice Research Institute, Los Baños, Philippines.   232

IRRI (International Rice Research Institute) (1967) *Annual Report*, International Rice Research Institute, Los Baños, Philippines.   72

IRRI (International Rich Research Institute) (1991) *World Rice Statistics*, International Rice Research Institute, Los Baños, Philippines.   67

Jenkinson, D. S. (1991) The Rothamsted long-term experiments: are they still of use, *Agron. J.*, **83**, 2–10.   86

Johansen, C. A., Mayer, D. F., Eves, J. D. and Kios, C. W. (1983) Pesticides and bees, *Env. Ent.*, **12**(5), 1513–1518.   218

Johnsen, C. A. and Mayer, D. F. (1990) *Pollinator Protection: a Bee and Pesticide Handbook*, Wicwes Press, Cheshire, CT.   218

Johnson, N. L. and Leone, F. C. (1976) *Statistics and Experimental Design in Engineering and in the Physical Sciences*, Vol. 2, 2nd Edn, Wiley, New York.   37

Johnston, A. E. (1974) Experiments made on Stackyard Field, Woburn, 1876–1974. *Rothamsted Experimental Station, Report for 1974, Part 2*, 29–44.

Johnston, A. E. (1994) The Rothamsted Classical Experiments, in Leigh, R. A. and Johnston, A. E. (eds) (1994) *Long-term Experiments in Agricultural and Ecological Sciences*, CAB International, Wallingford. 173

Johnston, A. E. and Jones, K. C. (1993) The Cadmium issue—long-term changes in the cadmium content of soils and crops grown on them, in Schulz, J. J. (ed) (1993) *Phosphate Fertilizers and the Environment*, International Fertilizer Development Centre, Muscle Shoals, USA. 205

Johnston, A. E. and Powlson, D. S. (1993) The Setting-up, Conduct and Applicability of Long-term Field Experiments in Agricultural Research, in Greenland, D. J. and Szabolcs, I. (1993) (eds) *Soil Resilience and Sustainable Land Use*, CAB International, Wallingford. 173

Johnston, A. E. and Powlson, D. S. (1994) The setting-up, conduct and applicability of long-term, continuing field experiments in agricultural research, in Greenland, D. J. and Szabolcs, I. (eds) (1994) *Soil Resilience and Sustainable Land Use*, 395–421. CAB International. 237, 242

Jorgenson, D. W., Gollop, F. M. and Fraumeni, B. M. (1987) *Productivity and U.S. Economic Growth*, Harvard University Press. 48

Kemmler, G. (1982) The need for long-term potassium fertilizer trials, Phosphorus and Potassium in the Tropics, in Pushparajah, E. and Sharifuddin, H. H. (eds) (1982) The Malaysian Society of Soil Science, Kaulalumpur. 150

Khush, G. S. and Coffman, W. R. (1977) Genetic Evaluation and Utilization (GEU), the rice improvement program at the International Rice Research Institute, *Theor. Appl. Genet.*, **51**, 97–110. 76

Koyama, T. and App, A. (1979) Nitrogen balance in flooded rice soils, *Nitrogen and rice*, pp. 95–103, International Rice Research Institute, Los Baños, Philippines. 77

Krauss, H. A. and Allmaras, R. A. (1982) Technology masks the effects of soil erosion on wheat yields—A case study in Whitman county Washington, in Schmidt B. L. *et al.* (eds) (1982) *Determinants of Soil Loss Tolerance, Spec. Publ.*, **45**. Am. Soc. Agron., Madison, WI. 97

Kropff, M. J., van Laar, H. H. and ten Berg, H. F. M. (1993) *ORYZA1: a basic model for lowland rice production*, International Rice Research Institute, Los Baños, Philippines. 79

Lanham, B. T., jr. (1947) *Farm Power and Equipment Costs in Northern Alabama*, Bulletin 260, Agricultural Experiment Station of the Alabama Polytechnic Institute, Auburn, Alabama. 51

Lanham, B. T., jr. and Lagrone, W. F., jr. (1942) *Increasing Incomes and Conserving Resources on Cotton Corn—Farms in Marion County, Alabama*, Bulletin 256, Agricultural Experiment Station of the Alabama Polytechnic Institute, Auburn, Alabama. 51

Larson, W. E., Pierce, F. J. and Dowdy, R. H. (1983) The threat of soil erosion to long-term crop production, *Science*, **219**, 458–465. 86

Lingard, J. and Rayner, A. J. (1975) Productivity Growth in Agriculture: A Measurement Framework, *Journal of Agricultural Economics*, **26**, 87–104. 25

Lloyd, T. A. (1992) *Present Value Models of Agriculture Land Prices in England and Wales*, unpublished Ph.D. Thesis, University of Nottingham. 180

Lu, Y., Cline, P. and Quance, L. (1979) *Prospects for Productivity Growth in U.S. Agriculture*, Agricultural Economic Report No. 435, Economics, Statistics and Cooperative Service, USDA, Washington, DC. 48

Lynam, J. K. and Herdt, R. W. (1989) Sense and sustainability: sustainability as an objective in international agriculture research. *J. Agric. Econ.*, **3**, 381–398. 5, 25, 64, 85, 112, 115, 172

Macnab, S., Stoltz, M., Tuck, B., Murphy, J. and Cross, T. (1988) Dryland wheat production and marketing costs in Oregon's Columbia Plateau, 1988–1989, *Ext. Serv. Circular*, Oregon State Univ., Corvallis, OR. 94

Madhun, Y. A. and Freed, V. H. (1990) Impact of pesticides on the environment, in Cheng, C. C. (ed) (1990) *Pesticides in the Soil Environment: Processes, Impacts and Modelling*, 429–464, Soil Science Society of America Book Series, #2, Soil Science Society of America, Madison. 217

MAFF (1866–1988) *Agricultural Statistics, United Kingdom 1866–1988*, HMSO, London. 186

MAFF (1968) *A Century of Agricultural Statistics, Great Britain 1866–1966*, HMSO, London. 180

Mahan, J. N. and Marsh, J. F. (1943) *Prices Received by Alabama Farmers for Farm Products, August 1909–August 1942*, Bulletin 258, Agricultural Experiment Station of the Alabama Polytechnic Institute, Auburn, Alabama. 51

Mann, H. H. and Tambane, V. A. (1910) The alkali lands of Niravally, *Bull. Dept. Agric.*, Bombay, 39. 151, 156

Marks, H. F. (1989) *A Hundred Years of British Food and Farming. A Statistical Survey*, London. 180

Mayton, E. L. and Roy, K. B. (1952) *Cotton–Dairy Farming in Alabama's Piedmont*, Circular 111, Agricultural Experiment Station of the Alabama Polytechnic Institue, Auburn, Alabama. 51

McCool, D. K., Brown, D. F., Papendick, R. I. and McDonough, L. M. (1987) Fate of herbicides on eastern Washington dryland: loss in runoff and movement and persistence in soil, in Elliott, F.L. (ed) (1987), *STEEP—Conservation Concepts and Accomplishments*, Washington State Univ., Pullman, WA. 97

Mew, T. (1991) Disease management in rice, in *CRC Handbook of Pest Management in Agriculture*, in Pimentel, D. (ed) (1991), Second Edition, Vol. III., pp. 279–299, CRC Press Inc., Boston. 82

Mineau, P. P. (1987) An evaluation of avian impact assessment techniques following broad-scale forest insecticide sprays, *Environ. Toxicol. Chem.*, **6**(10), 781–791. 221

Mitchell, C. C., Westerman, R. L., Brown, J. R. and Peck, T. R. (1991) Overview of long-term agronomic research. *Agron. J.*, **83**, 24–29. 86

Morrison, F. B. (1947) *Feeds and Feeding*, 20th edn, McGraw-Hill Publishing Co., New York, NY. 115

Morse, R. A. (1989) History of subsection Cb: apiculture and social insects, *Bulletin of the Entomological Society of America*, **35**(3), 115–119. 218

Mortenson, J. L. (1963) Complexing of metals by soil organic matter, *Soil Sci. Am. Proc.*, **27**, 179–186. 156

Motavalli, P. P., Bundy, L. G., Rudraski, T. W. and Peterson, A. E. (1992) Residual effects of long-term nitrogen fertilization on nitrogen availability to corn, *J. Prod. Agric.*, **5**, 363–368. 112

Nambiar, K. K. M. and Ghosh, A. B. (1984) Highlights of Research of Long-Term Fertilizer Experiments in India (1971–82), in *Long Term Fertilizer Experiments Research Bulletin No. 1*, Indian Agricultural Research Institute, New Delhi, India. 72, 148

Nambiar, K. K. M. (1985a) Response to potassium and soil balance studies in major cropping systems under long-term fertilizer use, in *Potassium in Agricultural Soils, Proc. of the Int. Sym.*, 9–10 March 1985, Soil Science Society of Bangladesh, Bangladesh Agricultural Research Council, Dhaka, India. 142, 148, 150

Nambiar, K. K. M. (1985b) All India Coordinated Research Project on Long Term Fertilizer Experiments and its research achievements, *Fertiliser News*, **30**(4), 56–66. 140, 141

Nambiar, K. K. M. (1986) All India Coordinated Research Project on Long Term Fertilizer Experiments (Indian Council of Agricultural Research), *Annual Progress Report 1984–85*, New Delhi, India. 165

Nambiar, K. K. M. (1988) Crop response to sulphur and sulphur balance in intensive

## References   251

cropping systems, in *Sulphur in Indian Agriculture, TSI-FAI Sym.* 9–11 March 1988, New Delhi, India.   146, 149, 150, 151, 155, 156

Nambiar, K. K. M. (1989) All India Coordinated Research Project on Long Term Fertilizer Experiments (Indian Council of Agricultural Research), *Annual Progress Report 1985–86 & 1986–87*, New Delhi, India.   138, 165, 166

Nambiar, K. K. M. (1990). Long-term fertility effects of wheat productivity, in Sanders, D. A. (ed) (1990) *Wheat for Nontraditional Warm Areas*, Mexico, D.F., CIMMYT (1991).   138, 143

Nambiar, K. K. M. (1991a) Sustainable agriculture: A challenging goal, *Process & Plant Engineering*, **IX**(3), 91–97.   138, 143, 156

Nambiar, K. K. M. (1991b) Integrating chemical fertilizers and organic manures for sustained productivity, in Somani, K. L., Totawat, K. L. and Baser, B. L. (eds) (1991) *Proceedings of National Seminar on Natural Farming*, Udaipur, India.   138, 150, 151, 156, 165

Nambiar, K. K. M. (1992a) Yield sustainability of rice–rice and rice–wheat systems under long-term fertilizer use, in Sadaphal, M. N. and Rajat, De. (eds) (1992) *Resource Management for Sustained Crop Production*, Indian Society of Agronomy, New Delhi, India.   150, 156, 158, 160, 163, 164, 165, 166

Nambiar, K. K. M. (1992b) All India Coordinated Research Project on Long Term Fertilizer Experiments (Indian Council of Agricultural Research), *Annual Progress Report 1987–88 & 1988–89*, New Delhi, India.   139, 141, 146, 147, 152, 153, 154, 167

Nambiar, K. K. M. and Abrol, I. P. (1988) Critical level of phosphorus in Vertisols in relation to crop responses, in Phosphorus in Indian Vertisols, *Summary Proceedings of a Workshop*, 23–26 Aug. 1988, ICRISAT Centre, India, Patancheru, AP-562324, India.   138, 143, 147, 149

Nambiar, K. K. M. and Abrol, I. P. (1989) Long Term Fertilizer Experiments in India—An Overview, *Fertiliser News*, **34**(4), 11–20 & 26.   145, 151, 156, 165

Nambiar, K. K. M. and Ghosh, A. B. (1985) Crop Response to Potassium and Soil Potassium balance in long-term fertilizer experiments, in *Soil Testing, Plant Analysis and Fertilizer Evaluation for Potassium, Proc. of a Group Discussion*, Nov. 22–23, 1985 (PRII Research Review Series 4).   148

Nambiar, K. K. M. and Ghosh, A. B. (1986) Nutrient balance in long-term experiments. *Potash Review*, Subject 27, No. 61986, Berne, Switzerland.   148

Nelson, D. W. and Sommers, L. E. (1972) A simple digestion procedure for estimation of total nitrogen in soils and sediments, *J. Environ. Qual.*, **1**, 423–425.   90

Nix, J. (1967, 1968, 1970, 1972, 1973, 1975, 1977 1992) *Wye College. University of London. The Farm Management Pocket Book*, 1st–22nd editions, Publication Department of Agricultural Economics, Ashford.   180, 181, 182, 183, 184, 185

Norguard R. (1991) *Sustainability: The Paradigmatic Challenge to Agricultural Economics*, Paper presented at 21st Conference of the International Association of Agricultural Economists, Aug. 22–29, Tokyo.   3

Nowak, P. J., Timmons, J., Carlson, J. and Miles, R. (1985) Economic and social perspectives on T values relative to soil erosion and crop productivity, in Follet, R. F. and Stewart, B. A. (eds) (1985) *Soil Erosion and Crop Productivity*, Am. Soc. Agron., Madison, WI.   86, 104

Oregon Dept. of Agriculture (1992) *1991–92 Oregon Agriculture & Fisheries Statistics*, Oregon Ag. Stat. Ser., Portland, OR.   107

Ou, S. H. (1985) *Rice Diseases*, Second Edition, CAB International, UK.   76

Oveson, M. M. and Besse, R. S. (1967) The Pendleton experiment station—its development, program, and accomplishments: 1928 to 1966. *Spec. Rept. 233*, Agric. Expt. Stn., Oregon State Univ., Corvallis, OR.   86

Painter, K. (1991) Does sustainable farming pay? *Journal of Sus. Ag.*, **1**(3), 37–48.   97, 106

Painter, K., Hinman, H., Miller, B. and Burns, J. (1991) 1991 enterprise budgets: eastern Whitman county, *Washington State Coop. Ex. Serv. Bull. EB1437*, Wash, St. Univ., Pullman, WA. 94

Painter, K. and Young. D. L. (1993) Alternative agricultural policies: environmental and economic trade-offs by policy 1993 field day proceedings: highlights of research progress. *Tech. Report 93–4*. Dept. of Crop and Soil Sciences, Wash. St. Univ., Pullman, WA. 97, 106

Penning de Vries, F. W. T., Jansen, D. M., ten Berge, H. F. M. and Bakema, A. (1989) *Simulation of Ecophysiological Processes of Growth in Several Annual Crops*, Pudoc Wageningen, Netherlands. 243

Phipps, T. T., Crosson, P. R. and Price, K. A. (1986) *Agriculture and the Environment*, Resources for the Future, Washington, DC. 225

Pimentel, D., Acquay, H., Biltonen, M., Rice, P., Silva, M., Nelson, J., Lipner, V., Giordano, S., Horowitz, A. and D'Amore, M. (1991) Assessment of environmental and economic impacts of pesticide use, Cornell University. 218, 219, 220, 221

Pimentel, D., Sandow, D., Dyson-Hudson, R., Gallahan, D., Jacobson, S., Irish, M., Kroop, S., Moss, A., Schreiner, I., Shepard, M., Thompson, T. and Vinzant, B. (1980) Environmental and social costs of pesticides: a preliminary assessment, *Oikos*, **34**, 127–140. 216, 217

Pingali, P. L., Marquez, C. and Palis, F. G. (1994) Pesticides and Philippine rice farmer health: medical and economic analysis, *Amer. J. Agric. Econo.*, **76**, 587–592. 84

Powlson, D. S. and Johnston, A. E. (1994) Long-term field experiments: their importance in understanding sustainable land use, in Greenland, D. J. and Szabolcs, I. (eds) (1994) *Soil Resilience and Sustainable Land Use*, 367–393. CAB International. 242

Rasmussen, P. E., Collins, H. P. and Smiley, R. W. (1989) Long-term management effects on soil productivity and crop yield in semi-arid regions of eastern Oregon, *Oregon Agric. Exp. Stn. Bull. 675*, Oregon St. Univ. and USDA-Agric. Res. Serv. 85, 96

Rasmussen, P. E. and Parton, W. J. (1994) Long-term effects of residue management in wheat/fallow: 1. Inputs crop yield and soil organic matter change, *Soil Sci. Soc. Am. J.*, **58**, 523–530. 88, 91, 92, 93

Rasmussen, P. E., Rickman, R. W. and Douglas, C. L. jr. (1986) Air and soil temperature changes during spring burning of standing wheat stubble, *Agron. J.*, **78**, 261–263. 89

Rasmussen, P. E., Allmaras, R. R., Rohde, C. R. and Roager, N. C. jr. (1980) Crop residue influences on soil carbon and nitrogen in a wheat-fallow system, *Soil Sci. Soc. Am. J.*, **44**, 596–600. 92

Rather, J. B. (1917) An accurate loss-on-ignition method for delineation of organic matter in soils, *Arkansas Agric. Exp. Stn. Bull*, 140. 90

Repetto, R. (1990) Environmental productivity and why it is so important, *Challenge*, September–October 1990. 223, 224

Ribaudo, M. O. (1989) *Water Quality Benefits from the Conservation Reserve Programme*, USDA, Resources and Technology Division, ERS, Agricultural Economic Report No. 606. 54, 226, 227, 228

Ribaudo, M. O. (1989) *Water Quality Benefits from the Conservation Reserve Programme. Agric. Econ. Rep*, 606. Res. and Tech. Div, USDA-Econ. Res. Serv., Washington, DC. 97

Ribaudo, M. O. (1989) *Water Quality Benefits from the Conservation Reserve Programme*, US Dept. of Agriculture, Resources and Technology Division, Economic Research Service, Agricultural Economic Report No. 606. 117

Robinson, R. W. (1951) *Cotton Production Practices in the Limestone Valley Areas of Alabama*, Circular 100, Agricultural Experiment Station of the Alabama Polytechnic Institute, Auburn, Alabama. 51

Rola, A. C. and Pingali, P. L. (1993) *Pesticides, rice productivity and farmers' health: an economic assessment*, World Resources Institute, Washington, DC and International Rice Research Institute, Los Baños, Philippines. 84

Rosegrant, M. W. and Pingali, P. L. (1994) Sustaining rice productivity growth in Asia: A policy perspective, in *J. International Development* (in press). 65

Rosenstock, L. (1992) Conceptual and methodological aspects of assessing the health impact of pesticides, paper presented at the Conference on *Pesticides, Environment and Health*, held at the Rockefeller Conference Centre, Bellagio, Italy. 217

Rothamsted Experimental Station (1855–1991) *Yields of the Field Experiments*, Rothamsted Experimental Station. 185

Rothamsted Experimental Station (1970) Rothamsted Experimental Station, Details of the Classical and Long-term Experiments up to 1967.

Ruttan, V. N. (1991) *Sustainable Growth in Agricultural Production: Poetry, Policy and Science*, CREDIT Research Paper No. 91/13, University of Nottingham. 23

Saito, M., and Watanabe, I. (1978). Organic matter production in rice field floodwater, *Soil Sci. Plant Nutrition*, **24**, 427–440. 77

Sibbesen, E. (1986) Soil movement in long-term field experiments, *Plant and Soil*, **91**, 73–85. 19

Sidhu, D. S. and Byerlee, D. (1992) *Technical Change and Wheat Productivity in the Indian-Punjab in the Post-Green Revolution Period*, CIMMYT Economics Working Paper 92–02, Mexico, D. F.: CIMMYT. 48

Steichen, J. M. (1979) *Estimating Soil Losses in Northern Missouri*, Extension Guide 1560, University of Missouri Extension Division, Columbia, MO. 117

Steiner, R. (1992) Personal communication, The Rockefeller Foundation, New York, NY. 117

Steiner, R., McLaughlin, L. (1992) Measuring the Externalities of U.S. Agriculture for Use in the Economic Analysis of Long-term Agronomic Experiments, *International Report*, Rockefeller Foundation, New York, NY. 199

Steiner R. A. and Herdt, R. W. (1995) *The Directory of Long-term Agricultural Experiments: Volume 1*, FAO, Rome. 16

Stevenson, J. H. (1989) Rothamsted: A Cradle of Agricultural Research, *Plants Today*, May–June 1989, 84–89. 173

Sutherland, G. J., Carlson, G. A. and Hoover, D. M. (1971) *Cost of Producing Cotton in the Southeast, 1966*, Economics Information Report No. 25, Dept. of Economics, North Carolina Experiment Station. 51

Sutherland, G. J., Carlson, G. A. and Hoover, D. M. (1974) *Distribution of Costs and Returns in Upland Cotton in Specific Subregions, Regions and the United States, 1969*, Economics Information Report No. 39, Dept of Economics, North Carolina Experiment Station. 51

Tabatabai, M. A. and Bremner, J. M. (1970) Use of the Leco automatic 70-second carbon analyzer for total carbon analysis of soils. *Soil Sci. Soc. Am. Proc.*, **34**, 608–610. 90

Tate, R. L. (1979) Effect of flooding on microbial activities in organic soils: carbon metabolism, *Soil Sci.*, **128**, 267–273. 77

Technicon (1976) Individual-simultaneous Determination of Nitrogen and/or Phosphorus in BD Acid Digests: *Industrial Method No. 334–74/A*, Technicon Industrial Systems, Tarrytown, NY. 90

Trimble, S. W. (1974) *Man-Induced Soil Erosion On The Southern Piedmont 1700–1970*, Soil Conservation Society of America, Ankeny, Iowa. 55

US Dept. of Agriculture (Issues 1899 through 1919) *Yearbook of the Department of Agriculture*, Government Printing Office, Washington, DC. 42, 51

US Dept. of Agriculture (1992) *Cotton and Wool Situation and Outlook Report*, Commodity & Economics Division, Economic Research Service, USDA, CWS-67. 49

US Dept. of Agriculture (Issue 1923–1925) *Agricultural Yearbook*, Government Printing Office, Washington, DC. 49

US Dept. of Agriculture (Issues 1920–1922) *U.S. Department of Agriculture Yearbook*, Government Printing Office, Washington, DC. 49

US Dept. of Agriculture (Issues 1927–1935) *Yearbook of Agriculture*, U.S. Government Printing Office, Washington, DC. 49

US Dept. of Agriculture (Issues 1944–1992) *Agricultural Prices*, Agricultural Statistics Board, National Agricultural Statistics Service, Washington, DC. 49

United States Department of Agriculture (1960) Prices paid by farmers for commodities and services: 1910–1960, *Stat. Bul. No. 319*, Stat. Rap. Serv. Washington, DC. 94

United States Department of Agriculture (1969) Agricultural prices: 1968 annual summary, *Stat. Rep. Serv. Pr1–3 (69)*, Washington, DC. 94

United States Department of Agriculture (1976) Index numbers of prices received and prices paid by farmers, *Stat. Rep. Serv. Pr1–3 (76)*, Washington, DC. 94

United States Department of Agriculture (1984) Agricultural prices: 1983 summary, *Nat. Agr. Stat. Serv. Pr1–3 (84)*, Washington, DC. 94

United States Department of Agriculture (1990) Agricultural prices: 1990 summary, *Nat. Agr. Stat. Serv. Pr1–3 (91)*, Washington, DC. 94

United States Department of Agriculture (1990) State-level costs of production: major field crops, 1987–89, *Econ. Res. Serv. Stat. Bul. No. 838*, Washington, DC. 94

United States Department of Agriculture (1991) Agricultural prices: 1991 summary, *Nat. Agr. Stat. Serv. Pr1–3 (92)*, Washington, DC. 94

United States Department of Agriculture (1992) Cost of Production—Major Field Crops, 1990, *Econ. Res. Serv.*, Washington, DC. 94

Upchurch, W. J., Kinder, R. J., Brown, J. R., and Wagner, G. H. (1985) *Sanborn Field: Historical Perspective*, Missouri Agricultural Experiment Station, Res. Bull. 1054, Columbia, MO. 111, 112

US Dept of Agriculture, *Cash Rents for Farms, Cropland and Pasture*, USDA Statistical Bulletin No. 813. 122

van Keulen, H. and Seligman, N. G. (1987) *Simulation of Water Use, Nitrogen Nutrition, and Growth of a Spring Wheat Crop*, Pudoc Wageningen, Netherlands. 243

Vaquer, A. (1984) La production algae dans les rizieres de Carargue pendant la periode de submersion, *Ver. Internat. Verein. Limnol.*, **22**, 1651–1654. 77

Wagner, G. H. (1989) Lessons in soil organic matter from Sanborn Field, in Brown, J. R. (ed) (1989) *Proceedings of the Sanborn Field Centennial*, Missouri Agricultural Experiment Station, Special Report 415, Univ. of Missouri, Columbia, MO. 112

Walker, D. J. and Young, D. J. (1986) The effect of technical progress on erosion damage and economic incentives for soil conservation. *Land Economics*, **62**, 83–93. 97

Watson, J. A. S. and More, J. A. (1924, 1933, 1941, 1944, 1945, 1949) *Agriculture: The Science and Practice of British Farming*, 1st, 3rd, 5th, 7th 8th editions, Edinburgh. 180, 182, 183

Webster, T. C. and Peng, Y. S. (1989) Short-term and long-term effects of methamidophos on brood rearing in honey bee colonies. *J. Econ. Ent.*, **82** (1), 69–74. 218

White, M. (1951) *Cotton Production Practices in the Piedmont Area of Alabama*, Circular No. 102, Agricultural Experiment Station of the Alabama Polytechnic Institute. Auburn, Alabama. 49, 51

Williams, J. R., Jones, C. A., Dyke, P. T. (1984) A modeling approach to determining the relationship between erosion and soil productivity, *Trans. ASAE*, **27**, 129–144. 117

Williams, J. R., Dyke, P. T., Fuchs, W. W., Benson, V. W., Rice, O. W. and Taylor, E. D. (1989) in Sharply, A. N. and Williams, J. R. (eds) (1989) EPIC-Erosion productivity Impact Calculator: 2. User Manual, U.S. Department of Agriculture Technical Bulletin No. 1768, Temple, Texas. 226

Williams, J. R. et al. (1990) *EPIC-Erosion Productivity Impact Calculator*, USDA, ARS, Technical Bulletin 1768, Temple, Texas. 55

Wischmeier, W. H. and Smith, D. D. (1978) *Predicting Rainfall Erosion Losses—a Guide to Conservation Planning*, Agricultural Handbook 537, U.S. Dept. of Agriculture, Washington, DC. 117

Woodhead, T., Huke, R. and Huke, E. (1993) Area, Locations, and ongoing collaborative research for the rice-wheat systems of Asia, *Workshop on Rice-Wheat Cropping*, June 1993, Food and Agriculture Organization, Rome. 232

World Commission on Environment and Development (1987) *Our Common Future*, OUP, Oxford. 1

Yamagichi, T., Okada, K., and Murata, Y. (1980) Cycling of carbon in a paddy field, *Jpn. J. Crop Sci.*, **49**, 135–145. 77

Yeager, J. H. (1968) *100 Years, Alabama Crop, Livestock and Income Data, 1866–1966*, Agricultural Economics and Rural Sociology, Agricultural Experiment Station, Auburn University, Alabama. 51

Yeager, J. H., Belcher, O. D. and Walkup, H. G. (1960) *Fertilizer Use and Practices by Alabama farmers*, Bulletin 320, Agricultural Experiment Station, Auburn University. 51

Young, D. L., Taylor, D. B. and Papendick, R. I. (1985) Separating erosion and technology impacts on winter wheat yields in the Palouse: a statistical approach, in McCool, D. K. (ed) (1985) *Erosion and Soil Productivity*, ASAE Publ. 8–85, Am. Soc. of Agric. Engr., St. Joseph, MI. 97, 105

# Index

activity of production  25
aggregate output/input index  30–2
  Fisher  31–2, 237
  Laspeyres  30–2, 38, 48, 187, 237
  Paasche  30–2, 187, 237
  Tornqvist–Theil  29, 31, 37, 121–31, 189, 237
aggregator function  27–9
Agricultural Gazette, The  183–4
Agricultural and Mechanical College of Alabama  42
Alabama Agricultural Experiment Station  42
Alabama Cooperative Extension Service  49, 51, 60
Alabama Old Rotation *see* Old Rotation
All-India Coordinated Research Project on Long Term Fertilizer Experiments  134
American Society of Agronomy, definition of sustainability  133–4
Andaqueptic Haplaquoll soil  78
animals
  farming in Midwest US  131
  killed by pesticides  221–2
Annual Abstract of Statistics (Central Statistical Office)  183, 185
Aomori Experiment Station, Japan  18
arithmetic indexes for Rothamsted Classicals  187–8, 190–3
Asia, rice yield and productivity trends  63–84
Auburn University, Alabama  41–2

basic slag  183–4
bee poisoning from pesticides  218–19
Bhubaneswar, Orissa, sustainability on laterite soil  134–5, 150–6
biological sustainability, compared with economic for wheat/fallow systems  99–104, 107
biological/physical dimension defined  7–8
birds killed by pesticides  221–2
blue baby syndrome  225
blue-green algae  77
boll weevil in 1915 on Alabama cotton  49–50
bone ash  183–4
boron toxicity  75, 76
Broadbalk
  A3 index  194–6
  cadmium  205
  changes in wheat yield  238–9
  crossed-factor experimental design  173
  output/input ratios  196, 198, 200–1
  replication  233
  Rothamsted Classicals  177–81
  wheat plots compared with IRRI  72
burning stubble in wheat/fallow systems  88–92

C-factor (cropping management factor)  117
cadmium
  Broadbalk and Park Grass  205
  Sanborn Field  111
cancer incidence among farmers  217
carbamate insecticides  221
carbon
  equivalents as common unit of measurement  10
  homeostasis of C and N in rice

carbon *(cont.)*
   soils 77
   soil organic (SOC)
      at Bhubaneswar, Orissa 152–3
      at Jabalpur, Madhya Pradesh 146–7
      at Ludhiana, Punjab 139–40
      at Pantnagar, Uttar Pradesh 159–62
      loss in wheat/fallow farming 92–3, 104
      in LT rice production 77–8
      and monocropping 229
      wheat yield 240–1
cation exchange capacity (CEC) 162
Central Luzon, LT rice yield experiment 67–84
Central Statistical Office, Annual Abstract of Statistics 183, 185
chain-linking index 32, 188
change in the state of technology 25–6
chemical control costs 211–13
China, area production and yield of rice 66
Classicals *see* Rothamsted Classicals
Claypan Farm and soil erosion 117
climate *see* weather
clover, crimson 42–4, red 113, 118, 130
Coastal Plain soil 44
Cobb–Douglas production function 28–9
Coimbatore 20
Columbia Basin Agricultural Research Center, Oregon 85–109
Columbia Plateau, erosion 97
Comer Agricultural Hall fire 49
Common Agricultural Policy, MacSharry Plan 171
corn, Sanborn Field 113, 116, 118, 130, *see also* maize
cost function, aggregator function 28–9
costs
   clean environment 211–15
   cotton production in Old Rotation 50
   fertilizer regulatory 224
   fish losses due to pesticides and manure 220–1
   honey-bee poisoning from pesticides 218–19
   horses or tractors at Rothamsted 182
   labour at Rothamsted 180–1
   off-site, of soil erosion 226–9
   pesticide externalities 222–4
   rented land at Rothamsted 180
   safe drinking water 215
   wheat/fallow systems 93–4
cotton, long-term experiment (Old Rotation) 41–61
cowpea 42–5, 47, 159–60
cropping systems
   C-factor (cropping management factor) 117
   changes in LT experiments 19
   erosion rates 226
   farm field 8–9
   irrigated rice in Asia 63–84
   linking structure to function 64–5
   long-term (LT) experiments and productivity indexes 231–44
   LT cotton experiment 41–61
   major, in India 133–69
   output/input efficiency 239–40
   summary of trends in productivity 234–5
Cullars Rotation experiment 49
CUSUM plot for shift in TFP index 37

$D$-value (transport coefficient of soil) 19
data collection and management 20–1
database, constructing aggregate output/input indexes 32
DDT insecticide 20, 54, 222
defoliants in Old Rotation 54
design of experiments *see* experimental design
disembodied technological change 25–6
Divisia index for Old Rotation 48
drinking water, nitrates in 97
Duggar, John F 42

economic dimension defined 7–8
economic environment, and wheat/fallow systems 92–104, 107
ecosystem 10–12, 24, *see also* externalities
embodied technological change 26
energy as common unit of measurement 10
energy crisis of 1970s 96
environmental costs, fertilizers 225–6

environmental productivity, and
    sustainability   209–10
Environmental Protection Agency (EPA)
    97, 215–18, 220, 224, 225
EPIC (Erosion Productivity Impact
    Calculator)   55, 117, 226
erodibility factor   117
Erosion Productivity Impact Calculator
    (EPIC)   55, 117, 226
erosion, soil
    after stubble burning   92
    associated with externalities   226–9
    at Claypan Farm   117
    Columbia Plateau   97
    ecosystem health   11
    in Old Rotation   54–7
    Revised Universal Soil Loss Equation
        (RUSLE)   97–9, 104–5
    in Rothamsted Classicals   199
    Sanborn Field   116–19
eutrophication of surface water   97, and
    fish losses   225–6
experimental design
    cropping systems in India   136–7
    programme for Sanborn Field   112–13
    recommendations for LT experiments
        242
    Rothamsted Classicals   173–5
    wheat/fallow systems   87
    *see also* statistical design
externalities
    cost of pesticides   210–24
    costs, incorporated into productivity
        209–30
    included in TFP for Rothamsted
        Classicals   198–206
    soil erosion and   226–9
    total social factor productivity (TSFP)
        and   9–10

FAO (Food and Agricultural
    Organisation)   15, 66, 133
farming
    cancer in farmers   217
    compared with LT experiments   18
    costs and wheat/fallow systems   94–7
    extrapolating trends from LT
        experiments   63–84
    in Midwest US   131

organic   227
productivity   4–5
rice yield and productivity trends
    66–71
Federal Insecticide, Fungicide and
    Rodenticide Act (FIFRA)   211–13
fertilizers
    at Bhubaneswar, Orissa   150–6
    at Broadbalk   177–9
    at Rothamsted Classicals   175–6, 183–4
    for cotton in Old Rotation experiment
        41–61
    environmental costs   225–6
    externalities   224–6
    fish losses   225
    health costs to society   224–5
    in Jabalpur, Madhya Pradesh   144–50
    in Ludhiana, Punjab   138–43
    Pantnagar, Uttar Pradesh   157–67
    Sanborn Field   114
    wheat/fallow systems   87–8
FIFRA (Federal Insecticide, Fungicide and
    Rodenticide Act)   211–13
fire at Comer Agricultural Hall   49
fish
    global food   3–4
    losses from eutrophication   225–6
    losses from pesticides   219–21
Fisher index   31–2, 237
fixed coefficient production function
    27–8
flowchart, sustainability   12
Food and Agricultural Organisation
    (FAO)   15, 66, 133
food, global   3–4, 133
fungicides
    externality costs and benefits   99
    Rothamsted Classicals   185

Gardener's Chronicle, The   183
genetic engineering   24
Gentlemen's Estate Book   183–4
geometric indexes for Rothamsted
    Classicals   189–90
geometric production function   28–9
ginning   51, 58
government programmes in US   96,
    105–7
grass, Rothamsted Classicals   173, 175–6

## Index

grassy stunt in LT rice production  76
green manure  227, *see also* manure
green revolution and LT experiments  20, 243

harvester, cotton  53, 61
hay at Park Grass  175, 185–6
health
  effects of pesticides  215–17
  fertilizers  224–5
  nitrate and nitrite contamination  224–5
herbicides
  birdlife and  221
  externality costs and benefits  99
  in Old Rotation  54
  Rothamsted Classicals  185
heterotrophic bacteria  77
honey-bee poisoning from pesticides  218–19
horses, Rothamsted Classicals  182–3

ICAR (Indian Council of Agricultural Research)  134, 138–9, 144–9
index
  A3 for Broadbalk  194–6
  A3 for Park Grass  191–3
  A3 for Woburn Continuous Wheat  193–5, 202
  aggregate input  27, 29, 31
  aggregate output to aggregate inputs  26, 30–2
  arithmetic, for Rothamsted Classicals  187–8, 190–3
  chaining  32
  cost equation for wheat/fallow systems  94
  difference in Laspeyres, Paasche, Fisher and Tornqvist–Theil  237
  Divisia, for Old Rotation  48
  Fisher  31–2, 237
  geometric, for Rothamsted Classicals  189–90
  Laspeyres  30–2, 38, 48, 187, 237
  Paasche  30–2, 187, 237
  productivity, for Old Rotation  47–8
  set at 100 for agronomical trial  33–7
  Tornqvist–Theil, of aggregate input  29, 31, 37, 121–31, 189, 237

*see also* output/input, total factor productivity (TFP) *and* total social factor productivity (TSFP)
India, area production and yield of rice  66–7
Indian Agricultural Research Institute, LT fertilizer experiments  72
Indian Council of Agricultural Research (ICAR)  144–9, 151–9
  fertilizer experiments  134, 138–9
  rice at Pantnagar, Uttar Pradesh  157–64
Indonesia, rice production  65
input index  26–7
insecticide
  carbamate and wildlife  221
  DDT  20, 54, 222
  drift  9
International Rice Research Institute  67, 71–5, 233

Jabalpur, Madhya Pradesh, sustainability on medium black soil  134–5, 144–50

$K$-factor of soil  117
kainit  45
kharif (monsoon) crop  136
Krishna, Andhra Pradesh, irrigated rice LT experiments  67–84

labour, Rothamsted Classicals  180–1
Laguna, LT rice yield experiment  67–84
Laspeyres index  30–2, 38, 237
  for Old Rotation  48
  for Rothamsted Classicals  187
Lawes, John Bennet  172–3
legumes, winter, in Old Rotation  41–4, 47, 53, 57–60
Leontief production function  27–8
lime, Rothamsted Classicals  183–4
loess overlying basalt  86
Long-Term Continuous Cropping Experiment (LTCCE), at IRRI Research Farm, Laguna, Philippines  72–83, 233
Long-Term Fertility Experiment, at PhilRice Research Station, Central Luzon  73–5

Index 261

long-term (LT) experiments
   analysis of productivity trends   241–3
   changes and reliability   19–20
   data   20–1
   farming compared with   4–5, 18
   geographical breakdown   16
   irrigated rice in Asia   63–84
   maintaining relevance   232–6
   quantitative measures of sustainability in   236–9
   Rothamsted Classicals   172–206
   sites   15–17
   soil archive   20
   soil movement   18–19
   statistical design limitations   17
   summary   234–5
   and sustainability of cropping systems   231–44
Ludhiana, Punjab, sustainability on alluvial soils   134–5, 137–43

McGruder plots, USA   18
machinery
   effect of cotton harvester in Old Rotation   61
   in Old Rotation   51
   Rothamsted Classicals   182–3
MacSharry Plan, Common Agricultural Policy   171
maize, crops at Ludhiana, Punjab   137–43, *see also* corn
manure
   barnyard for wheat/fallow systems   87
   farmyard
      on Broadbalk plots   198, 203–4
      in Jabalpur, Madhya Pradesh   150
      in Ludhiana, Punjab   143
      in Pantnagar, Uttar Pradesh   157–9
      Rothamsted Classicals   177–9, 181, 183–4
      on Sanborn Field   111
   fish losses due to   220–1
   green   227
   organic matter (OM) in wheat/fallow systems   104–5
   pea vines   19, for wheat/fallow systems   87, 90
   Sanborn Field   114–16
Marvyn loamy sand   44, 49

mesic Typic Haploxerolls (Walla Walla silt loam)   86
methemoglobinemia (blue baby syndrome)   225
model
   development of simulation   243
   EPIC for soil erosion   55
   trend of TFP index   36–7
monocropping and carbon (organic matter)   229
monsoon rice at Bhubaneswar, Orissa   150–2
Morrow plots, USA   18
mules in Old Rotation   51
multi-crop systems   25, comparisons, Sanborn Fields   111–32
muriate of potash   45

neutral technical change   26, 27
Nitram   178, 183
nitrate contamination   224–5
nitrites and health hazards   225
nitrogen
   and soil organic carbon   241
   chemical nitrogen   57–60
   cropping systems in India   136–7, 140
   homeostasis of C and N in rice soils   77
   N factor productivity and yield decline in rice   66, 75–83
   N response experiments by IRRI   74–5
   no N fertilizer for cotton in Old Rotation   41–61,
   red clover   130
   Sanborn Fields   111, 114–15
   in soil at Bhubaneswar, Orissa   153
   in soil at Jabalpur, Madhya Pradesh   147
   in soil at Ludhiana, Punjab   138–43
   in soil at Pantnagar, Uttar Pradesh   162–3
   in wheat/fallow systems   87–8, 90–2
   trends in rice yield in India   68–9
Norfolk fine sandy loam   44

oat crop, Sanborn Field   113, 118
Old Permanent Manurial experiments in India, size of   17

Old Rotation
  acid soil infertility  232–3
  changes in yield  239
  conclusions  61
  data  49–51
  experiment at Auburn University
      41–2
  historic perspective  42
  productivity analysis  47–8
  results  51–60
  site characteristics and soil
      measurements  42–7
  treatments  42
  TSFP indexes  54–7
  winter legume  41–4, 53
    chemical nitrogen  57–60
organic matter (OM), content in
    wheat/fallow systems  104–5, 115,
    see also manure
organophosphates and wildlife  221
output/input
  index of aggregate  26, 30–2
  ratio
    Broadbalk plots  196, 198, 200–1
    Park Grass  193, 197
    for Sanborn Field  121–5
    Woburn Continuous Wheat  194–5,
      200–2
    see also total factor productivity (TFP)
    and total social factor productivity
    (TSFP)$_f$

Paasche index  30–2, 187, 237
Pacolet sandy clay loam  44, 55
Pantnagar, Uttar Pradesh
  cowpea in  159–60
  rice-wheat long term experiment  232
  soil and climate  134–5
  soil organic carbon  159–62
  sustainability on foothill (tarai) soil
      156–67
Park Grass
  A3 index  191–3
  cadmium in herbiage  205
  hay at  175, 185–6
  output/input ratios  193, 197
  Rothamsted Classicals  173–6
  split-plot factorial experiment  173
pea vines  19, 87, 90

pedological characteristics of a plot  24
pesticides
  bee poisoning  218–19
  crop and tree loss  218
  effect in Rothamsted Classicals  202
  environment externalities of  210–24
  externality costs and benefits  97–9
  fish loss  219–21
  human health  215–17
  in Old Rotation  54
  pest control strategies  20
  Rothamsted Classicals  185
  wildlife killed by  221–2
pH values
  at Ludhiana, Punjab  140
  at Pantnagar, Uttar Pradesh  158–60
  cropping system  11
  in Old Rotation  46–7
Philippines, LT rice yield experiment
    67–84
phosphorus
  cropping systems in Index  136–42
  depletion in soil  10
  eutrophication of surface water  97
  fertilization in Old Rotation  53
  in Old Rotation  44–6
  Sanborn Field  111, 114–15
  in soil at Bhubaneswar, Orissa  153–4
  in soil at Jabalpur, Madhya Pradesh
      147–8
  in soil at Ludhiana, Punjab  141
  in soil at Pantnagar, Uttar Pradesh
      163–4
  soil movement  18–19
Piedmont soils  44
plot sizes, LT experiments  18–19
pollution  9
  control costs  211–15
  industrial growth  24
population, global  3–4
  productivity  4, 244
potassium
  cropping systems in India  136–42
  fertilization in Old Rotation  44–6, 53
  Sanborn Field  114–15
  in soil at Bhubaneswar, Orissa  154–5
  in soil at Jabalpur, Madhya Pradesh
      148, 150
  in soil at Ludhiana, Punjab  141–2

in soil at Pantnagar, Uttar Pradesh 164–5
prices
  in aggregate output/input indexes 30–1
  as common unit of measurement 10, 30
  fertilizer for Rothamsted Classicals 183–4
  index of, and wheat 94–7
  seed for Rothamsted Classicals 184
  wheat/fallow systems 105–7
  *see also* costs
production function 25, 27–8, 29–30, 70
productivity
  analysis for Old Rotation 47–8
  cotton in Old Rotation 51–60
  declining, and degradation in soil 244
  environmental, and sustainability 209–10
  externality costs 209–30
  growth 26, 30–1
  partial 26
  pesticide benefits 97–9
    costs 222–4
  population growth 4–5, 244
  recommendations for LT experiments 241–3
  for Sanborn Field multi-crop 120–32
  summary 234–5
  and sustainability of cropping systems 231–44
  trends in rice systems 64–84
  *see also* total factor productivity (TFP) *and* total social factor productivity (TSFP)
profitability ratio 32–3
profits, wheat/fallow systems 105–7

rahi (winter) crop 136
red clover, Sanborn Field 113, 118, 130
references 245–55
reliability of data in LT experiments 20–1
replication, non-replication in Old Rotation 42–4
  *see also* experimental design
Residue Management Experiment, Oregon 240–1, 243

Revised Universal Soil Loss Equation (RUSLE) 97–9, 104–5
rhizosphere 77
rice
  at Pantnagar, Uttar Pradesh 156–67
  irrigated, systems in Asia 63–84
  regional and national trends in Asia 65–6
  wild 4
  yield trend at Bhubaneswar, Orissa 150–6
rock phosphate 114
rotation
  at Jabalpur, Madhya Pradesh 149–50
  at Pantnagar, Uttar Pradesh 159–67
  corn at Sanborn Field 125, 130
Rothamsted Experimental Station
  Classicals, long-term (LT) experiments 171–206
  *see also* Broadbalk, Park Grass *and* Woburn Continuous Wheat
Royal Agricultural Society of England and continuous wheat experiment 176–7

Safe Drinking Water Act (SDWA) 215
Sanborn Fields, multi-crop comparisons 111–32
SDWA (Safe Drinking Water Act) 215
seed prices for Rothamsted Classicals 184
sheath blight in LT rice production 75, 80
silage drainage and fish losses 220
sites, long-term (LT) experiments 15–17
size of plots in LT experiments 18–19
SOC (soil organic carbon) *see* carbon, soil organic
social dimension defined 7–8
soil
  alluvial, at Ludhiana, Punjab 137–43
  climate 134–5
  analysis in wheat/fallow systems 89–90, 92
  characteristics for LT experiments 21
  characteristics in Sanborn Field 114
  cropping systems in India 134–5, 139–43
  $D$-value (transport coefficient of soil) 19

soil (cont.)
  erodibility factor   117
  erosion
    after stubble burning   92
    associated with externalities   226–9
    at Claypan Farm   117
    Columbia Plateau   97
    ecosystem health   10–12
    in Old Rotation   54–7
    Revised Universal Soil Loss
      Equation (RUSLE)   97–9, 104–5
    in Rothamsted Classicals   199
    Sanborn Field   116–19
  fertility as capital stock   24
  foothill (tarai), at Pantnagar, Uttar
    Pradesh   134–5, 156–67
  laterite, at Bhubaneswar, Orissa
    134–5, 150–6
  measurements and types for Old
    Rotation   42–7
  medium black, at Jabalpur, Madhya
    Pradesh   134–5, 144–50
  movement in LT experiments   18–19
  organic carbon (SOC) *see* carbon, soil
    organic
  organic-C loss in wheat/fallow
    farming   104
  phosphorus depletion *see* phosphorus
  quality enhancement   227–9
  quality for Rothamsted Classicals
    203–6
  samples in LT rice production   76–9
  system performance and, quality
    239–41
Southern Piedmont of the United States
  erosion   55
Soviet Union, grain embargo against   96
soy bean, at Jabalpur, Madhya Pradesh
  144–6, 149–50
statistical design
  assessment of TFP indexes   35–7
  limitations of LT experiments   17
  Old Rotation   42–4
  Rothamsted Classicals   17, 172–206
  *see also* experimental design
stem rot in LT rice production   75, 80
straw burning   89–92
sulphur in soil
  at Bhubaneswar, Orissa   151, 155–6

  at Jabalpur, Madhya Pradesh   149, 150
  at Pantnagar, Uttar Pradesh   165–7
sun as a dying star   6
sustainability
  alluvial soil at Ludhiana, Punjab
    137–43
  biological/physical dimension defined
    7–8
  concepts   3–13
  definition
    for agronomic trial   33
    by the American Society of
      Agronomy   133–4
    by FAO   133
    economic dimension   7–8
    Lynam and Herdt   5, 112
    for Rothamsted Classicals   171–2,
      191
    wheat/fallow system   85
  economic, compared with biological
    99–104
  ecosystem health   10–12
  erosion effects   97
  flowchart   12
  on foothill (tarai) soil at Pantnagar,
    Uttar Pradesh   156–67
  future research needs   169
  on laterite soil at Bhubaneswar, Orissa
    150–6
  measurement   8–12
  medium black soils at Jabalpur,
    Madhya Pradesh   144–50
  quantitative measures of, in long-term
    (LT) experiments   236–9
  Rothamsted experience   171–206
  social dimension defined   7–8
  solutions for biological, in
    wheat/fallow systems   108–9
  spatial levels   5–6
tarai soil   156–67
technology of production   25–7
  *see also* total factor productivity (TFP)
  *and* total social factor productivity
    (TSFP)
TFP *see* total factor productivity (TFP)
thermic Typic Hapludults   44
time, sustainability and   6–7
time paths for TFP   34–5

# Index

Tornqvist–Theil index  237
  of aggregate input  29, 31, 37, 189
  for Sanborn Field multi-crop  121–31
total factor productivity (TFP)
  advantages and disadvantages  237
  changes in rice yield  82–4
  definition, measure of productivity  9, 64–5
  including externalities for Rothamsted Classicals  198–206
  index for Old Rotation  48, 52–4, 56
  indexes for Rothamsted Classicals  179, 186–98
  measurement of index  23–37
  Sanborn Field  121–31
  trend and variability  33–5
  trends in rice yield in India  70
total social factor productivity (TSFP)
  compared with TFP  10
  ecosystem health  12–13
  index  9–10
  LT rice experiments in Asia  84
  Old Rotation experiments  54–8
  Rothamsted Classicals  172, 174, 179
  wheat/fallow systems  86, 92–107
tractors in Old Rotation  51
translogarithmic production function  29–30
tree loss and pesticides  218
TSFP *see* total social factor productivity (TSFP)
tungro virus disease in LT rice production  76

Universal Soil Loss Equation (USLE)  117

vetch  42–4, 47
viral diseases in LT rice production  75, 76

water
  drinking, nitrates in  97
  nitrate contamination  224–5
  pollution  213–15
  safe drinking  213–15
weather
  at Bhubaneswar, Orissa  150–1
  at Jabalpur, Madhya Pradesh  150
  Columbia Plateau  86
  Indian experimental sites  134

Sanborn Field  120–1, 124, 131
wheat
  at Broadbalk  174–206
  at Jabalpur, Madhya Pradesh  145–6, 150
  at Ludhiana, Punjab  137–43
  at Pantnagar, Uttar Pradesh  156–67
  at Woburn Continuous Wheat experiment  176–7
  changes in yields at Broadbalk  238–9
  Sanborn Field  113, 116, 118
  seed prices for Rothamsted Classicals  184
  wheat/fallow systems in Pacific NW America  85–109
  yield and soil organic carbon  240–1
wildlife and pesticides  221–2
Woburn Continuous Wheat
  A3 index  193–5, 202
  experiment  174–206
  output/input ratios  194–5, 198, 200–2
World Bank and LT experiments  15

yield
  at Bhubaneswar, Orissa  150–2
  causes of decline in rice  75–83
  changes in wheat, at Broadbalk  238–9
  cotton in Old Rotation  50, 52–60
  cropping systems in India  138–9
  defined  8
  hay at Park Grass  175, 185–6
  history of wheat  90–92
  measurement of sustainability and  8
  negative trends  243–4
  Philippines rice  71
  rice at Pantnagar, Uttar Pradesh  156–67
  Rothamsted Classicals  179, 185–6
  Sanborn Field  118–30
  trends, rice in Asia  65–83
  TSFP and  9
  wheat and soy bean at Jabalpur, Madhya Pradesh  144–6, 149–50
  in wheat/fallow systems  89

zinc
  deficiency  75
  farmyard manure and  142–3

zinc *(cont.)*
   in soil at Jabalpur, Madhya Pradesh 150
   in soil at Pantnagar, Uttar Pradesh 165–7

*Index compiled by Dr. M.P.M. Merrington*